GLASS FIBRE REINFORCED CEMENT

GLASS FIBRE REINFORCED CEMENT

A. J. Majumdar, OBE, PhD

and

V. Laws, MSc

*Published on behalf of the
Building Research Establishment*

OXFORD
BSP PROFESSIONAL BOOKS
LONDON EDINBURGH BOSTON
MELBOURNE PARIS BERLIN VIENNA

Copyright © (British) Crown copyright 1991. Published by permission of the Controller of Her Britannic Majesty's Stationery Office

BSP Professional Books
A division of Blackwell Scientific Publications Ltd
Editorial offices:
Osney Mead, Oxford OX2 0EL
25 John Street, London WC1N 2BL
23 Ainslie Place, Edinburgh EH3 6AJ
3 Cambridge Center, Cambridge, MA 02142, USA
54 University Street, Carlton, Victoria 3053, Australia

All rights reserved. No part of this publication may be reproduced, stored in a retrieval system, or transmitted in any form or by any means, electronic, mechanical, photocopying, recording or otherwise without the prior permission of the copyright owner.

First published 1991

Set by Setrite Typesetters
Printed and bound in Great Britain by Hartnolls, Bodmin, Cornwall

DISTRIBUTORS

Marston Book Services Ltd
PO Box 87
Oxford OX2 0DT
(*Orders*: Tel: 0865 791155
        Fax: 0865 791927
        Telex: 837515)

USA
Blackwell Scientific Publications, Inc.
3 Cambridge Center
Cambridge, MA 02142
(*Orders*: Tel: (800) 759-6102)

Canada
Oxford University Press
70 Wynford Drive
Don Mills
Ontario M3C 1J9
(*Orders*: Tel: (416) 441-2941)

Australia
Blackwell Scientific Publications
(Australia) Pty Ltd
54 University Street
Carlton, Victoria 3053
(*Orders*: Tel: (03) 347-0300)

British Library
Cataloguing in Publication Data
Majumdar, A. J.
  Glass fibre reinforced cement.
  1. Construction materials. Glass fibre reinforced cement
  I. Title   II. Laws, V.   III. Building Research Establishment
  624.1833

  ISBN 0-632-02904-8

Acknowledgements

The authors wish to thank Dr B A Proctor (now retired from Pilkington Brothers plc) for his helpful comments on the original manuscript, and their colleagues at BRE for their cooperation, particularly P L Walton for preparing the index.

Contents

Foreword *by Dr I. Dunstan*		vii
1	**Alkali-Resistant Glass Fibres**	**1**
	Historical	1
	The matrix phase	5
	Alkali-resistant glass fibres	10
	Glass/cement interactions	17
2	**Theoretical Principles**	**26**
	Introduction	26
	Notation	27
	Mechanism of reinforcement: aligned long fibre composites	28
	Effect of fibre length and orientation: efficiency factors	34
	Fibre/cement bond	49
3	**Production Methods for Grc Components**	**55**
	Constituent materials	55
	Spray production methods	56
	Mix and place methods	61
	Other (miscellaneous) processes	62
	Curing	66
	Surface finishes	67
	Quality control	67
4	**Properties of Portland Cement Grc**	**70**
	Spray-dewatered grc	70
	Premixed grc	86
	Other properties of grc	87
5	**Grc From Modified Portland Cement Matrices**	**92**
	Fillers	92
	Pozzolanas	93
	Lightweight grc	107

6	**Polymer Modified Grc**	**112**
	Polymer modified AR glass fibre reinforced cement	113
	Polymer modified E-glass fibre reinforced cement	125
	Polymer modified E-glass reinforced high-alumina cement	128
	Polymer impregnated grc	129
7	**Non-Portland Cement Grc**	**130**
	High-alumina cement (HAC) composites	131
	Supersulphated cement (SSC) composites	134
	Portland blastfurnace cement (PBFC) composites	139
	Other cement composites	139
	Glass fibre reinforced autoclaved calcium silicate (grcs)	140
8	**Microstructure of Grc and Glass/Cement Bond**	**143**
	Initial microstructure	143
	Microstructure changes with time	144
	Microstructure of Cem-FIL 2/grc	150
	Microstructure of grc made from non-Portland and blended cements	150
	Microstructure and bond	154
	Microstructure and cracking	161
9	**Durability**	**164**
10	**Applications and Future Developments**	**177**
	Applications	178
	Future prospects	181
References		183
Index		193

Foreword

Glass fibre reinforced cement, grc for short, is an interesting construction material of considerable potential. The development of its potential was one of the most exciting research projects being carried out by the Building Research Establishment during my time as Director between 1979 and 1983.

To make the material, a small amount of an alkali-resistant glass fibre is incorporated in cement mortar to overcome the traditional weakness of inorganic cements, namely poor tensile strength and brittleness. In grc the length and content of the glass fibre reinforcement can be chosen to meet the strength and toughness requirements of the product. Manufacturing methods have been specially developed for producing articles of various shapes and designs, from architectural cladding panels to housing for electrical machines, and the material is playing an important part in the renovation of old buildings. Conceptually, grc is similar to asbestos cement and as the use of the latter is declining in some countries the use of grc as a substitute is increasing. This trend is likely to grow.

The first commercial alkali-resistant glass fibre, Cem-FIL, was produced in the UK by Pilkington Brothers PLC in 1971 after pioneering work on glass compositions had been carried out at the Building Research Establishment. Prior to this a considerable amount of research and development work on grc was done in the USSR in the late 1950s and early 1960s using ordinary glass fibres. But it is fair to say that the production of the AR glass fibre Cem-FIL, by the collaborative efforts of Pilkington, the National Research Development Corporation (now part of the British Technology Group) and BRE, marked the beginning of the grc industry. AR glass fibres similar to Cem-FIL have now been produced in the USA and in Japan. On a worldwide basis the grc industry is currently supplying goods and services worth several hundred million US dollars.

In the 1970s research on grc was generously supported at BRE. Work was done on fibre incorporation techniques, on different types of cement matrices (including Portland cement modified by various pozzolanic materials and polymers), on the fundamental chemistry of glass/cement interactions, and on some theoretical aspects of reinforcement of brittle matrices such as cements by glass fibres. One important aspect of this programme was to expose test boards made from different types of grc to different environments for assessment of their long-term properties. In 1979 Pilkington launched their second generation AR glass fibre Cem-FIL 2 provided with an 'inhibitive' surface coating that is considered to be much more resistant to corrosion by cements than the earlier Cem-FIL. At BRE, only a limited amount of work was undertaken on Cem-FIL 2; in the 1980s much of BRE's effort went towards accumulating results on the long-term properties of various types of the grc composites that were already under investigation.

The authors of this book have been involved in grc research at BRE from the very beginning and they have given a detailed account of research results obtained by them and their colleagues over a period spanning two decades. Results for up to ten years' storage in different environments are given for many grc composites, and in some cases longer-term results have been obtained. These results are supplemented by important contributions from other sources, notably from Pilkington. The production methods and current applications of grc are also described briefly.

The book is likely to be of interest to engineers, architects, specifiers and component manufacturers who wish to know about the development of various properties in the different types of grc over long periods of time in different environments. It will be of general interest to materials scientists in the construction field, not least because grc is an interesting example of a composite material where the components are reactive towards each other. The reader may learn how this inherent reactiveness has been counteracted to produce a material that is useful for many purposes. Above all, the book is a historical document of the development of the subject. The authors present their views about likely future developments, whilst being careful to point out the limitations of the material.

GRC has reached a stage where Standards are already in existence for several products made of it, and for test methods. Several other Standards are in preparation and I am sure that this book will be a significant contribution to that work.

Dr I. Dunstan
Director General,
British Standards Institution

Chapter One
Alkali-Resistant Glass Fibres

1.1 Historical

As the name implies, glass fibre reinforced cement (grc) comprises two essential components, namely glass fibres and cement. Inorganic hydraulic cements used in the construction industry are highly alkaline and therefore require special alkali-resistant glass fibres for their reinforcement. The development of such fibres during the past 20 years has been responsible for the foundation and growth of a new industry, the grc industry.

The art of drawing fibres from heat-softened glass was known to glass makers of ancient Egypt but the use of strong glass fibres as reinforcements is a relatively modern idea, materialising after the almost simultaneous development in the USA in the 1930s of (a) manufacturing methods for continuous glass fibres and (b) thermosetting polyester resins. Borosilicate E-glass fibres, originally developed for making electrical insulation tape, were used to manufacture strong and tough resin composites that could be formed to shape. This heralded the dawn of the glass fibre reinforced plastics (grp) industry.

Following the success of the grp industry in the 1950s attempts were made to examine the potential of glass fibres as reinforcements for concrete. Cement and concrete are weak in tension and against impact. Conventionally the weaknesses are overcome in concrete structures by the use of steel reinforcement bars or cables. One way to use glass fibres was to make up ropes of glass rovings joined together and coated with plastics so as to make a reinforcing bar or cable directly replacing steel. E-glass fibres were used and experiments were carried out with unstressed as well as stressed reinforcement[1,2]. The modulus of the glass is rather low for the former method to be efficient, but the combination of low modulus and high tensile strength makes stressed reinforcement a promising approach. However, very little interest has been shown in this area in recent years. Another and more successful approach has been to produce grc sheets, conceptually similar to asbestos cement, which can be moulded to shape. Most of the grc products now in use are made from thin sheets.

In these products cement is rarely used alone; a proportion of sand or other fine aggregates is needed to prevent cracking due to drying shrinkage. It is also now becoming common practice to modify the cement matrix by the addition of pozzolanas, admixtures or polymers to make it more suitable for reinforcement by fibres. In grc, such a matrix based on Portland and other cements is reinforced by a relatively small amount of glass fibres.

The first systematic study on grc as a composite material was carried out in the USSR by Biryukovich and his colleagues[3,4]. Borosilicate E-glass was used in these investigations, and as these are not resistant to alkalis special low-alkali cements or cement plus polymer were selected as the matrix. The reinforcement was used in a variety of forms including woven mats, and glass contents of up to 50% by weight of the finished product were achieved. Advantages claimed were high tensile strength and a high degree of elasticity in thin sections, water tightness and good thermal, acoustic and dielectric properties. Some prototype structures such as thin shell roofs were constructed.

An English translation of the Russian work became available in 1965[4]. This more or less coincided with the publication of pioneering research by Romualdi and Batson[5] on the reinforcement of concrete by steel wires and by Krenchel[6] on several types of fibre composite including fibre reinforced cement. Much interest was being directed at that time to metal matrix composites and to grp, and the fundamental basis of the science and technology of fibre composite materials was being laid. At the Building Research Establishment (BRE) a research project was launched in 1966 to look into the prospects of reinforcing cement and concretes with fibres. One of the objects of the project was to identify fibres that could be used as replacements for asbestos.

From the very outset the BRE work concentrated on glass fibres. It was soon realised that the approach taken by Biryukovich and co-workers of using special cements in combination with E-glass fibres would have only limited application and an alkali-resistant fibre was required for the reinforcement of Portland cements. Initially attempts were made to produce continuous crystalline fibres from glass by using the glass-ceramic approach of controlled nucleation and crystallisation. Fibres produced in this way were found to have very poor tensile strength arising from surface flaws created from faster crystallisation on the surface than in the bulk. It also became clear that the high speed of continuous fibre production is not compatible with uniform volume crystallisation of the glass. Attention was then focused on finding glass fibre compositions that were inherently more alkali-resistant than, say, E-glass fibres and examining their potential for cement reinforcement.

After a careful search of the literature it was concluded that glasses in the system $Na_2O-SiO_2-ZrO_2$ studied by Dimbleby and Turner[7] might possess the required degree of alkali resistance for reinforcing Portland cements. A glass with the composition (by weight) Na_2O 11%, SiO_2 71%, ZrO_2 16%, Al_2O_3 1%, Li_2O 1% and referred to as G20 was being produced in the USA by Corning Glass Works for use in boiler gauges and laboratory ware. Fibre from this glass was produced at BRE using a single tip bushing, and its performance in a Portland cement extract solution was studied and compared with that of E-glass fibres (Fig. 1.1(a))[8]. Subsequently a batch of

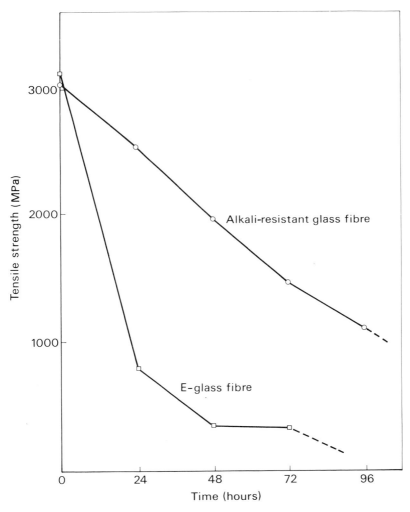

Fig. 1.1(a) Tensile strength of glass fibres in the aqueous solution phase of Portland cement at 80°C.

the G20 glass was obtained from Corning Glass Works and fiberised in the UK in the multifilament strand form. Portland cement composites made from this fibre were kept in different environments together with similar specimens made from E-glass fibres, and their strength properties were determined at various time intervals[9]. These results (Fig. 1.1(b)) clearly indicated that fibres made from glasses in the system $Na_2O-SiO_2-ZrO_2$ are more suitable for reinforcing cements than E-glass fibres. At about the same time work on zirconia containing glass fibres was also in progress in the USSR[10].

These preliminary results on the performance of an alkali-resistant (AR) glass fibre as cement reinforcement were of immediate interest to the glass industry in the UK, and one of the major glass producers in the world, Pilkington Brothers PLC, entered into a collaborative agreement with BRE and the National Research Development Corporation (now part of the British Technology Group) to develop a commercially viable AR glass fibre. Such a fibre, Trademarked Cem-FIL, was launched in 1971 by Fibreglass Limited, then a subsidiary of Pilkington. A few other AR glass fibres have been produced commercially since then. The properties of Cem-FIL fibres are compared with those of E-glass in Table 1.1.

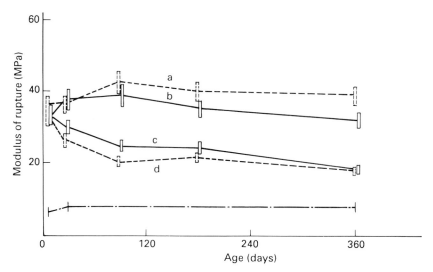

Fig. 1.1(b) Relation between modulus of rupture and age of glass fibre reinforced Portland cement composites. Fibre length 34 mm. Fibre content 4% by volume. (a) and (b) Alkali-resistant glass; (c) and (d) E-glass; ———, water cured at 18°C; ---, air (40% RH) cured at 18°C; —·—·—, average matrix strength, air and water cured.

Table 1.1 Glass fibre properties.

	E-glass[a]	AR glass[b]
Liquidus temperature (°C)	1140	1185
Fibre working temperature (°C)	1250	1283[c]
Young's modulus of fibre at 25°C (GPa)	75.5	70
Tensile strength of single fibre at 25°C (MPa)	3600	3600
Strain at failure	~2%	~2%
Tensile strength of fibre strand (MPa)	1700–2700	1450–1900
Density (Mg/m^3)	2.53	2.68
Coefficient of linear thermal expansion (per °C × 10^6)	5	7.5
Refractive index	1.550	1.561

a Lowenstein (1973)[20]
b Data from Pilkington Research and Development Laboratory
c Corresponding to 100 Pa s viscosity

With the commercial availability of Cem-FIL AR glass fibres the interest in grc as a new material of construction was immediate and its impact on the international construction scene has been sufficiently strong for an Association, named the Glass Fibre Reinforced Cement Association (GRCA) to have been formed in 1975. Through its Technical Committee the GRCA issue Codes of Practice and Guidelines for testing grc products. In safeguarding the interests of the young industry the GRCA is providing a valuable service for the producers and users of this new product.

It is worth pointing out that several applications of E-glass reinforced high-alumina cement composites were made in various countries before the advent of Cem-FIL AR glass fibres. It is believed that in many of these applications the grc material has performed adequately.

1.2 The matrix phase

In most of the applications of grc to date the proportion of glass fibres in the composite has been small — 5% by volume or less. The matrix phase is therefore very important in the development of composite properties and the durability of the material.

Various types of hydraulic cement — for example, Portland, high-alumina, supersulphated, Portland-blastfurnace slag, regulated set, etc. — are used in the construction industry, and in principle grc products can be made from any of them. However, as Portland cements are by far the most important of the inorganic cements, accounting for more than 90% of the world production of cements, nearly all grc products use this type of cement as the matrix at the present time. In recent years the use of blended cements has grown. In these a part of the Portland cement is replaced by a pozzolanic material such as pulverised fuel ash (pfa), ground granulated blastfurnace slag (ggbs) or microsilica (silica fume). The pozzolanas play an important role in combining with $Ca(OH)_2$ and other alkalis produced in cement hydration, thereby reducing the potential for corrosive alkaline attack on the glass fibres. It has also been found useful to add polymer dispersions of certain types — styrene–butadiene and acrylics are good examples — to control the flow properties of the slurry during the manufacture of grc components. By incorporating a low-density component such as perlite or pfa cenospheres, lightweight grc products have been made. Alternatively, a lightweight matrix can be produced by autoclaving lime plus silica or cement plus silica mixtures.

The properties of the matrix phase in grc are greatly influenced by the fabrication variables. The constitution and fineness of the cement powder and the amount of water used are the most important variables controlling the degree of hydration of the cement at a given time. A high water/cement (w/c) ratio produces a matrix that is very porous and therefore weak. On the other hand, if a low w/c ratio is used in grc manufacture much of the cement may remain unhydrated for a very long time when exposed to weather.

The effect of w/c and other fabrication variables on the properties of important hydraulic cements has been studied in detail and the results can be found in standard works of reference[11,12]. Since Portland cement is currently the principal binder in grc manufacture some of its important physical and engineering properties are listed in Table 1.2 as a guide. As the properties of the matrix are very dependent on w/c and curing conditions, a range of possible values is listed.

Table 1.2 Properties of hardened Portland cement paste.

Elastic modulus (GPa)	7 – 28
Compressive strength (MPa)	14 – 140
Tensile strength (MPa)	1.4 – 7
Modulus of rupture (MPa)	2.8 – 14
Tensile failure strain	0.02 – 0.06%
Poisson ratio	0.23 – 0.30
Thermal expansion (per °C × 10^6)	12 – 20
Volume change on drying	0.2 – 0.3% (negative)
Density (Mg/m^3)	1.7 – 2.2

The property of the matrix phase that is of utmost importance in grc is its alkalinity. It is not easy to obtain a realistic measure of this property at various stages in the life of the product, as the hydration of cement is a progressive process that can continue for many years. The presence of external components such as pozzolanas or polymers makes the attainment of equilibrium in hydration reactions even more difficult. The alkalinity of the matrix must refer to the composition of the pore solution that can be obtained by forcing out the solution from hardened cement pastes by the application of pressure. The technique, first used by Longuet et al.[13] is proving very useful and a considerable body of research results now exists on the compositions of pore solutions present in Portland and blended cement pastes and mortars. Nixon and Page[14] have recently reviewed the current state-of-the-art in this field with particular reference to alkali–aggregate reactions, and the following observations are taken from that review.

The alkalinity of the pore solution in cement pastes as measured by the hydroxyl ion concentration depends primarily on the amounts of the alkali metal oxides present in the cement. This is shown in Fig. 1.2 for 28 day old pastes studied by various investigators. There is not much information yet on the compositions of pore solutions in pastes that are several years old but Longuet et al.[13] found that the hydroxyl, sodium and potassium ion concentrations increased to a maximum between 7 and 28 days and then remained constant or decreased slightly. The calcium concentration fell to less than the detection limit in a few months. For Portland and other similar cements $Ca(OH)_2$ is a major product of hydration reactions. The lowest alkalinity of such a matrix would therefore correspond to that of a saturated $Ca(OH)_2$ solution, and this is shown in Fig. 1.2.

Pozzolanic additions such as microsilica (silica fume), metakaolin and pulverised fuel ash (pfa) to Portland cements are playing an increasingly important role in the grc industry, and the potential of glassy slags either in granulated or pelletised form is also being explored. Of these, microsilica additions have the largest effect in reducing the hydroxyl ion concentration in the cement pore solutions, the degree of reduction depending on the amount of microsilica added. The experimental results of Page and Vennesland[15] are shown in Fig. 1.3. Nixon and Page[14] conclude that pf ashes need a minimum of 28 days to have a significant effect on the pore solution chemistry and they reduce the hydroxyl ion concentrations more effectively in high-alkali cement pastes by means of pozzolanic reactions. In contrast, the available evidence suggests that granulated blastfurnace slags operate more as a diluent. Some typical values of the reduction of OH^- ion concentrations in the pore solutions in blended cement pastes are given in Table 1.3 from the work of Page and co-workers[16].

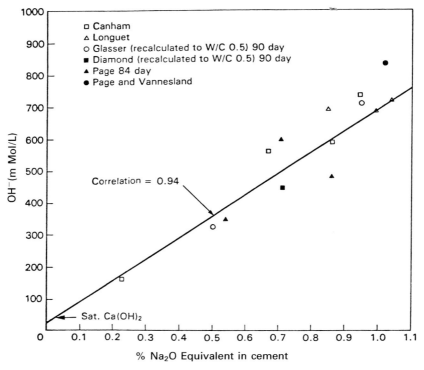

Fig. 1.2 Hydroxyl ion concentrations in pore solution of pastes with 0.5 w/c, 28 day results except where otherwise stated; approx. 20°C storage (Reference 14).

Table 1.3 Hydroxyl ion concentration in cement paste pore solution.

% Na_2O equivalent		wt% addition	$\dfrac{water}{solids}$ ratio	Storage time (days)	OH^- ion concentration (m mol/ℓ)
Cement	Addition				
0.67	—	—	0.45	28 365	604.0 661.7
0.67	pfa 3.37	40	0.45	28 365	331.1 319.6
0.67	slag 0.97	40	0.45	28 365	466.4 505.8
0.67	slag 0.97	60	0.45	28 365	341.1 358.9

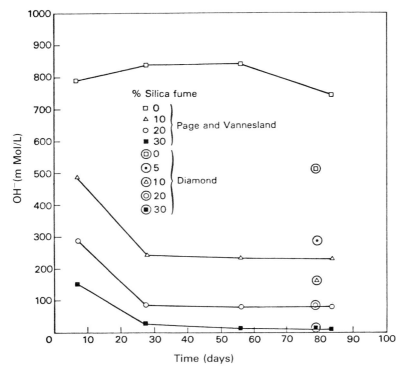

Fig. 1.3 Change in hydroxyl ion concentration with time and proportion of silica fume (Reference 15).

Alternative but less relevant methods of assessing cement alkalinity consist of suspending the ground cement paste in a known quantity of water or making a cement slurry with a very high w/c and keeping it stirred for a long period of time. The analysis of the solution phase gives some idea of the alkalinity of the matrix. The results from one series of experiments are shown in Table 1.4[17].

One recent innovation in the grc industry has been the use of a special type of cement in Japan and China. The cement manufactured by Chichibu Cement Company in Japan consists mainly of calcium silicates, $3CaO.SiO_2$ and $2CaO.SiO_2$, calcium aluminosulphate $3CaO.3Al_2O_3.CaSO_4$, calcium sulphate $CaSO_4$ and water quenched blastfurnace slag[18]. On hydration, the major products are calcium silicate hydrates (C–S–H) and ettringite $3CaO.Al_2O_3.3CaSO_4.32H_2O$. Very little, if any, calcium hydroxide is pre-

Table 1.4 Composition of the aqueous phase in contact with various cements.

g/litre	Portland cement	60% Portland cement + 40% pfa	60% Portland cement + 40% pozzolana*	High-alumina cement
Ca^{2+}	0.71	0.53	0.51	0.16
Na^+	0.24	0.33	0.28	0.06
K^+	0.57	0.56	0.54	0.30
Alkalinity as $(OH)^-$	1.04	0.95	0.87	0.67
pH	12.07	12.65	12.60	12.05

* Italian pozzolana from Naples.

sent in the hydrated cement and the OH^- ion concentration is low. Similar cements are now made in China for use in the grc industry. The pH value of the pore liquid in the low-alkalinity sulphoaluminate cement paste is around 10.5[19]. A sulphoaluminate cement has been in production in the UK also for a number of years.

1.3 Alkali-resistant glass fibres

Inorganic silicate glasses are inherently reactive towards alkalis; the silicon–oxygen–silicon network which forms the main structural skeleton of silicate glasses is attacked by the hydroxyl ions:

$$-\overset{|}{\underset{|}{Si}}-O-\overset{|}{\underset{|}{Si}}- + OH^- \rightarrow -\overset{|}{\underset{|}{Si}}-OH + -\overset{|}{\underset{|}{Si}}O^-$$
(in solution)

The addition of ZrO_2, which is present in all commercially available AR glass fibre compositions, reduces the rate of the silicate network breakdown in a substantial way but does not eliminate it altogether. Hence AR glass fibres are not immune to alkali attack from cements. Attempts to improve the alkali resistance of glass fibres are still continuing and it is instructive to discuss some of the factors that are important in the development of a new fibre.

1.3.1 Fiberising considerations

Continuous glass fibres are produced by the rapid mechanical attenuation of a stream of molten glass exuding under gravity through small orifices or tips located in the base of a rectangular electrically heated vessel made of platinum alloy and called a 'bushing'. A bushing is provided with a large number of tips, usually 200 or a multiple thereof, and produces the same number of filaments. The filaments are gathered together in the form of 'strands' and several strands are collected together to form a 'roving'. Loewenstein[20] has described all aspects of glass fibre production technology in detail.

For economic reasons, in commercial fibre drawing the working temperature T_w is preferably not allowed to exceed 1300°C and T_w should be at least 40–50°C above the liquidous temperature T_L[21]. The glass viscosity is maintained in a narrow range close to 100 Pa s for smooth running of the drawing operation. Although the alkali resistance of glasses increases with an increase in the proportion of ZrO_2 present, high zirconia glasses are more difficult to melt due to both a reduced rate of solution for the more refractory component and to a slower homogenisation of the melt at increasing melt viscosities. Phase separation and crystal growth in the glasses may create further difficulties. In formulating commercial compositions the manufacturers of AR glass fibres have tried to optimise the melting and fiberising characteristics of the glass without sacrificing their inherent alkali resistance. The viscosity–temperature relations of two zirconia-containing AR glasses are reproduced in Fig. 1.4 from the work of Proctor and Yale[21]. Both compositions were deemed to be suitable from the point of view of fibre production as T_w and $T_w - T_L$ values satisfied the operational requirements but the composition referred to as earlier glass produced an unacceptable level of crystal growth below T_L and hence was discarded.

Attempts are being made to produce glass fibres in the ZrO_2–SiO_2 system by the sol-gel process[22]. If successful such a process may circumvent the difficulties of producing high zirconia glass fibres by the conventional melt drawing method. The deposition of a suitable alkali-resistant coating on the glass fibre by the sol-gel route is also being explored.

1.3.2 Glass compositions

Numerous patents for AR glass fibre compositions that are claimed to be suitable for cement reinforcement have been granted, but only a few of them have reached the stage of commercial production even on a pilot scale. In addition to ZrO_2 it is believed that rare earth oxides such as La_2O_3 and CeO_2, and other oxides such as SnO_2, MnO, ZnO, Cr_2O_3, ThO_2 and TiO_2, impart sufficient alkali resistance to alkali silicate glasses to be of commercial interest. The most recent trend suggests the incorporation of a limited amount of rare earth oxides in the glass. It is too early to say whether these new compositions are economically viable. The influence of glass composition on the alkali resistance of fibres drawn from them has recently been reviewed by Majumdar[23] and by Fyles *et al.*[24] A few commer-

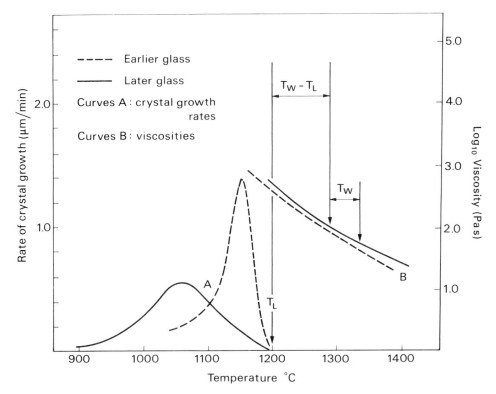

Fig. 1.4 Rates of crystal growth and viscosity against temperature (Reference 21).

cially important AR glass fibre compositions are listed in Table 1.5. For comparison the composition of E-glass fibres is included.

The alkali resistance of glass fibres is related to the extent of network breakdown in the glass when the fibres are reacted with an alkaline solution. Majumdar and colleagues[23] compared the extent of the network breakdown in borosilicate E-glass and typical zirconosilicate glasses used as AR glass fibres. A Portland cement extract solution was used as the reacting medium and experiments were carried out at three different temperatures. Some of the results are summarised in Fig. 1.5. The superior resistance of AR glass over E-glass to Portland cement extracts can be easily seen.

1.3.3 Coating

A coating or 'size', usually of an organic polymer applied to glass fibre during manufacture, is an essential component of the reinforcement pro-

Table 1.5 Alkali-resistant glass fibre compositions (wt%).

	E-glass	G20	Cem-FIL	Minelon	NEG	Asahi super	Chinese	AR slag wool (Chinese)
SiO_2	54.0	71.0	62.0	63.2	60.2	56.4	60.0	31.3
ZrO_2	—	16.0	16.7	14.3	19.8	16.9	14.5	10.0
TiO_2	—	—	0.1	—	—	—	6.0	11.3
Al_2O_3	15.0	1.0	0.8	1.6	0.3	—	—	13.4
Fe_2O_3	0.3	—	—	0.3	—	—	—	4.6
B_2O_3	7	—	—	—	—	—	—	—
Rare earth oxides	—	—	—	—	—	10.3	—	—
CaO	22.0	—	5.6	6.9	0.5	—	4.5	16.8
MgO	0.5	—	—	0.1	—	—	—	8.2
MnO_2	—	—	—	—	—	—	—	2.6
Na_2O	0.3	11.0	14.8	12.2	16.4	15.3	12.5	0.7
K_2O	0.8	—	—	0.3	2.2	0.9	2.5	0.1
Li_2O	—	1.0	—	—	1.0	—	—	—
F_2	0.3	—	—	—	—	—	—	—

duct. In addition to protecting the fibres from abrasion by handling equipment, the size also binds the individual filaments into convenient bundles for incorporation in the composite. Many attempts have been made in the recent past to coat glass fibres with alkali-resistant resins, but in commercial practice such a procedure has not been found to be particularly useful. Uneconomically high coating levels are needed to produce significant effects, and the presence of pinholes or cracking often renders the coating ineffective.

A novel approach in this respect has been the development of the surface treatment used for Cem-FIL 2 AR glass fibres[25]. In this process a chemical inhibitor is incorporated in the size, and this is slowly released in the alkaline environment of the set cement around the glass fibres. This results in a significant reduction in glass/cement interaction leading to a marked decrease in fibre strength loss in the cement environment[24]. Organic compounds of the polyhydroxy phenol family have been found to be very effective as chemical inhibitors[26]. The use of inhibitors in the size formulations for AR glass fibres is likely to be an important area of research in the future.

Alkali-resistant glass fibres such as Cem-FIL used in cement reinforcement usually have diameters larger than 10 μm. In cement composites the fibre reinforcements are not likely to be shorter than, say, 3 mm in length. These dimensions are greatly in excess of the 2.5 μm and 80 μm upper limits for diameter and length, respectively, that are considered to be particularly hazardous for fibres[27] in relation to the risk of producing lung diseases.

14 *Glass Fibre Reinforced Cement*

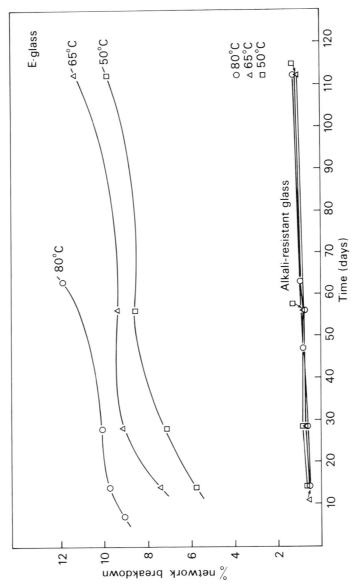

Fig. 1.5 Network breakdown of E-glass and alkali-resistant glass G20 in Portland cement extract solution.

1.3.4 Alkali-resistance tests

For screening new glass fibre compositions and for the quality control of existing commercial fibre it is necessary to adopt a procedure for assessing the alkali resistance of the fibre. This property is by far the most important in determining the suitability of the fibre for cement reinforcement.

In their original work on AR glass fibre compositions Majumdar and Ryder[8] measured the tensile strength of single glass filaments before and after they were reacted with (a) N/1 NaOH at 100°C for $1\frac{1}{2}$ h, (b) saturated $Ca(OH)_2$ at 100°C for 4 h and (c) cement aqueous phase solution at 80°, 50° and 25°C. The composition of the solution in (c) at room temperature (NaOH 0.88 g/ℓ, KOH 3.45 g/ℓ and $Ca(OH)_2$ 0.48 g/ℓ, pH 12.5) was chosen from the work of Lawrence[28]. The diameters of the fibres were also measured before and after the experiments. The decrease in the strength and diameter of the fibre after the tests was used to compare the alkali resistance of various glass compositions. These results were compared with those obtained with experimental composites made from cement using these fibres. Subsequently Majumdar and colleagues[23] studied glass/cement interactions employing extracts from cements used in composite fabrication as the reacting medium.

Over the years several other methods have been developed for testing the alkali resistance of glass fibres. For assessing the suitability of glass compositions for cement reinforcement, Proctor and Yale[21] modified the British Standard test for water resistance of glasses. Glass grains passing a 425 μm sieve but held by a 300 μm sieve were autoclaved at 121°C in contact with various alkaline solutions, e.g. 1 M and 0.1 M NaOH solutions. After a stipulated period of reaction time the extract solutions were analysed for glass components. The alkali resistance of various glasses in the system $Na_2O-SiO_2-ZrO_2-CaO$ was determined in this way.

As the search for glass fibres with improved alkali resistance continued it became clear that none of the tests described above simulate the environment of a cement composite. The reinforcement in cement is a multifilament fibre strand and it is in intimate contact with the solid phase in the composite. Changes in the glass fibre strand in the cement composite do not always correlate with the results of weight loss, chemical extractions or microscopic tests. Glass fibre strengths are critically dependent on stress concentrations on the surface arising from flaws, and the behaviour of multifilament strands in this respect cannot be easily deduced from results obtained with single fibres. The strand in cement (SIC) test developed by Litherland and his colleagues[29] overcomes most of the difficulties and is proving to be a very convenient and direct method for assessing the suitability of glass fibres for cement reinforcement. The method has already been adopted by the Glass

Fibre Reinforced Cement Association and may be incorporated in national and international Standards in future.

In the SIC test the tensile strength of a glass fibre strand placed in a matrix of hardened cement paste is determined directly. The test is carried out with specimens in which, as shown in Fig. 1.6, a block of cement surrounds the central portion of a glass fibre strand. Beyond the test length the strand is impregnated with resin to provide protection and extra strengthening of the fibre. The small plasticine grommets are positioned as illustrated in order to prevent adhesion between the cement and the resin.

After the placing, the cement is allowed to set and cure for 24 h at 20–25°C and 90–100% RH. The specimen is then transferred to the desired environment for a stipulated period and removed prior to testing. The specimen is loaded to failure in a tensile testing machine, gripping the resin impregnated ends of the strand in the jaws. The tensile strength of the strand is easily calculated from the breaking load and values of the relevant fibre parameters. For a valid result the strand should break inside the cement block.

For assessing the alkali resistance of a glass fibre, the SIC strength is usually measured immediately after a 24 h hardening period and then at three or more subsequent stages of an accelerated ageing schedule at two or more temperatures, commonly 50°C and 80°C.

SIC strength data obtained by Proctor and his colleagues[30] for several AR glass compositions having different amounts of ZrO_2 are shown in Fig. 1.7. It is easily seen that the proportion of strength retained in the fibre at 50°C increases with the increase in the amount of ZrO_2 in the glass up to about 16.8 wt% ZrO_2 and then levels off. The improved alkali resistance of Cem-FIL 2 fibres compared to Cem-FIL, brought about by the novel surface treatment in the former, is illustrated in Fig. 1.8 by the SIC strength data obtained for both types of fibre.

Glass fibres that are more alkali-resistant than Cem-FIL 2 have been identified and SIC test results on several prototype fibres have recently

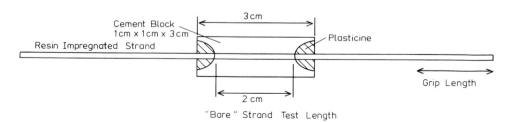

Fig. 1.6 SIC test specimen (Reference 29).

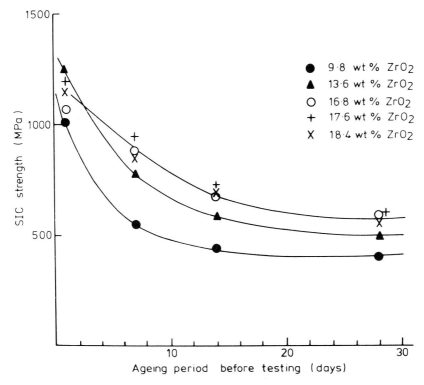

Fig. 1.7 Strengths of different zirconia glasses in set Portland cement, maintained wet at 50°C (Reference 30).

been reported by Fyles *et al.*[24]. It is likely that one of these fibres will be produced commercially in the near future.

1.4 Glass/cement interactions

It is clear from the discussion in the preceding sections that when placed in an alkaline environment such as that provided by hydrating Portland cements AR glass fibres will enter into chemical interactions with the medium. It is important to know what these interactions are and in particular the effects they may have on the strength of the glass fibre reinforcement.

Larner *et al.*[31] have studied the reactions between uncoated single filaments (~ 10 μm diameter) drawn from G20 glass (see Table 1.5) and two Portland cement extract solutions at different temperatures and for various lengths of time. After the initial filtration and collection of ions in the hydroxyl extracts, the fibre surfaces were successively treated with 2N acetic acid and

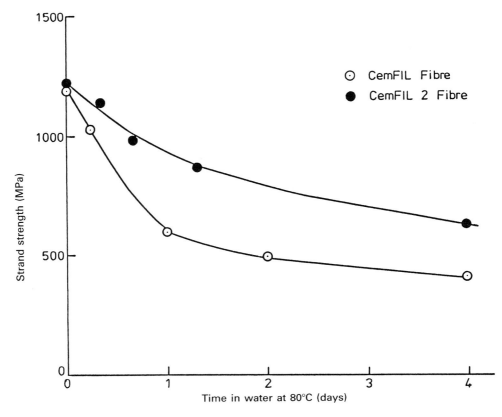

Fig. 1.8 Strand in cement strengths at 80°C (Proctor, *Proc. Int. GRC Congr.*, Paris, 1980).

50% HCl solutions. At 20°C changes in the pH and OH^- ion concentrations of the two cement extract solutions from their initial values (Table 1.6) brought about by reactions with the glass fibre over a period of two years were small, 0.7, 0.48 g/ℓ and 0.3, 0.25 g/ℓ, respectively, for the low- and high-alkali cements, and the reacting solutions remained highly alkaline.

Table 1.6 Typical analyses of final aqueous Portland cement extracts (g/ℓ of extract or g/100g of glass fibre).

	Batch no. 737	Batch no. 739
pH	12.7	12.9
Ca^{2+}	0.85	0.39
Na^+	0.26	0.34
K^+	0.56	2.15
$(SO_4)^{2-}$	0.69	0.65
$(OH)^-$	0.81	1.30

Fig. 1.9, giving the amounts of sodium removed from G20 glass fibres after reaction with the low-alkali cement extract solution, shows a rapidly decreasing rate of attack which tended to reduce almost to zero at very low levels of Na^+ removal. From Arrhenius-type plots made from the data in Fig. 1.9 a value of 18.5 kcal/mole is obtained for the energy of activation for the sodium removal process. This value is close to the 17.8−18.3 kcal/mole range quoted in the literature[32] for self-diffusion of Na^+. It is also very close to the energy of activation value of 18.9 ± 1.6 kcal/mole deter-

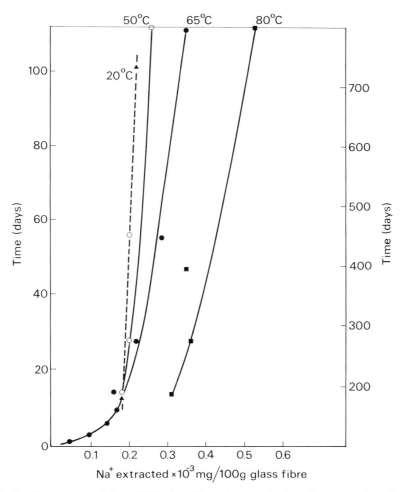

Fig. 1.9 Sodium extracted from G20 glass fibres into the hydroxyl extract after digestion in aqueous cement extract − OPC Batch No. 737 at 20, 50, 65 and 80°C. Ordinates on left-hand side refer to 50, 65 and 80°C; ordinates on right-hand side refer to 20°C.

mined by Langford et al.[33] for Na^+ depletion supporting the inter-diffusion model of Doremaus[34] for the hydration of soda glass. Although Larner et al.[31] did not obtain a $\sqrt{(time)}$ dependence of Na^+ removal that is consistent with the diffusion mechanism, later work by Simhan[35] on glass fibres containing 17 wt% ZrO_2 (Cem-FIL), and 9 wt% ZrO_2 reacted with water and cement aqueous phase composition of Lawrence[28], indicated $\sqrt{(time)}$ dependence for both Na and Si depletion in the two media studied.

The study by Larner et al. indicated that individual glass components from AR fibres were not leached to the same extent when reacted with cement extract solutions, nor did they all form alkali-soluble species. Calcium was completely removed from the cement extracts and deposited on the fibre either as $Ca(OH)_2$ or as part of a more complex product. The soluble reaction product was largely composed of calcium and silicon with a small amount of sodium, potassium and zirconium. There were also traces of lithium and aluminium which were present in the original glass. The lime/silica ratios of the reaction products suggested the possible formation of a poorly crystallised calcium silicate hydrate (C–S–H) gel. The current evidence suggests that much of the silicon and virtually all of the zirconium that is removed from AR glass fibres after reacting with the highly alkaline solutions present in hydrated Portland cements remain in the vicinity of the glass for a long time as part of the original glass fibre or as an insoluble reaction product.

An examination by electron spectroscopy (ESCA) of reacted G20 fibres before and after acid washing has shown[31] that the surface zirconium concentration was much greater after the reaction. This finding is in agreement with the energy dispersive X-ray analysis (EDXA) of the eroded glass fibre in a scanning electron microscope (SEM). In one experiment the Zr:Si ratio showed an increase from 0.11 to 0.26 as a result of the chemical attack by the cement extract solution, falling back to 0.17 after subsequent acid treatment. The precise nature of the zirconium bearing phase left on the surface could not be ascertained, but when it was removed by acid washing with 5N HCl the analysis of the extract indicated a $Na_2O:ZrO_2$ ratio of between 3 and 4. Other workers, for instance Simhan[35], have suggested that the reaction layer on the AR fibre may be that of hydrated zirconium. When AR glass fibre G20 was digested in N/1 NaOH at 50°C for 100 days the extreme form of alkaline attack produced the microstructure shown in Fig. 1.10. It is apparent that a reaction layer had formed on the surface of the fibre which subsequently detached itself from the bulk, possibly due to a mismatch in thermal expansion.

The chemical interactions between AR glass fibres such as Cem-FIL and cement pastes are more difficult to follow. The SEM/EDXA examination of the surface of fibres seen in broken grc samples have not revealed any

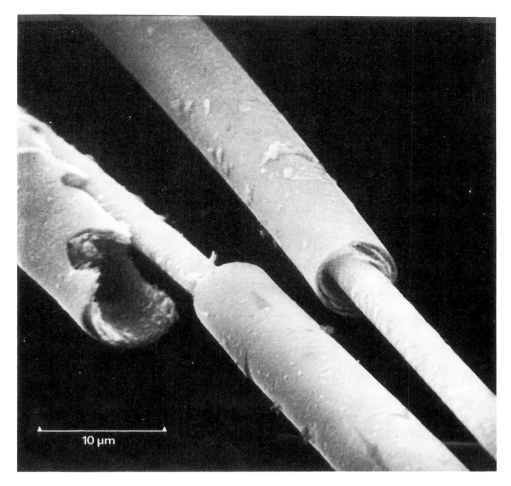

Fig. 1.10 G20 fibre after reaction with N/1 NaOH solution at 50°C.

significant difference in the analysis on fibres from aged composites from that of the pristine glass fibre, but the presence of Zr in the matrix was indicated in some instances[36]. It has also been claimed[37] that some Zr is to be found in large $Ca(OH)_2$ crystals. Working with bulk glasses of zirconosilicate compositions Chakroborty *et al.*[38] have shown by electron probe microanalysis (EPMA) that enrichment of Zr relative to Si takes place when the glass is reacted with Portland cement pastes. The formation of a film on the fibre surface containing Zr^{4+}, Ti^{4+} and Fe^{3+} has also been proposed by Li *et al.*[39] for their AR slag wool. The present state-of-the-art suggests that when a zirconium containing AR glass fibre is placed in an alkaline medium, the Zr−O bonds are only slightly attacked by OH^- ions compared to Si–O bonds. The increase in the relative concentration of Zr on the surface would have the effect of progressively reducing the area of the surface available for further attack. The process may eventually cease and the whole surface be effectively covered by zirconium ions or a zirconium-rich reaction product. This hypothesis is consistent with a logarithmic relationship for the rate of reaction similar to that found by Vernon et al.[40] for the oxidation of zinc in pure air. It is interesting to note that the rates of silica and soda removal from the G20 glass fibres by the cement extract solutions used by Larner et al.[31] best fitted logarithmic relationships. However, the energy of activation values for Na^+ depletion mentioned earlier and the work of Simhan[35] do not support this hypothesis.

From the chemical extraction data such as those shown in Fig. 1.9 it is not possible to speculate on the strength of the fibre resulting from the interaction. Glass fibre strengths are governed by the size and population of flaws on the surface, and if these flaws are made larger or more of them are formed because of the interaction it is perhaps justifiable to conclude that there will be a reduction in the strength of the fibre. This may be the case with fibre/cement-extract interactions at high temperatures. At ambient temperatures it is sometimes possible that some flaws on a glass fibre surface are etched away by reactions with a suitable solution producing an increase in fibre breaking load for a period of time. It is as yet unclear whether such a process occurs in the case of AR glass/cement-extract interactions. However, some unpublished results at BRE indicate that individual uncoated G20 filaments, carefully handled at all stages, lost very little of their tensile breaking load when kept for up to a year in a sealed polythene bag at 20°C in contact with various alkaline solutions (including a cement extract solution) having pH values up to 13. By contrast, at 50°C there was a considerable reduction in the strength of the glass filaments when reacted with solutions of pH > 11 for a period of only a few months. On the other hand there is some evidence[41] also that the strength of AR glass fibre strands is reduced significantly in alkaline solutions at ambient temperatures during the first three months.

If one examines the results obtained with filaments that were not so gently handled, e.g. in the experiments of Larner et al.[31] and as would be the case in the manufacture of a grc component, a slightly different picture emerges. Using the uncoated G20 filaments from their reaction rate studies[31] at 20°C and 65°C, Majumdar et al.[42] reported the results shown in Fig. 1.11. Two typical OPC extracts were used (Table 1.6). After digestion and removal of the bulk of the extract the fibres were treated with 2N acetic acid and washed with water, followed by a further treatment with 5N HCℓ acid. Finally, the fibres were washed with water and dried at room temperature and stored in screw-capped bottles prior to testing.

It is clear from Fig. 1.11(a) that when glass fibre was kept in continuous contact with Portland cement extracts at ambient temperature, its strength after the first six months was significantly lower than the initial strength. However, there is a firm indication that no further reduction in strength occurred up to two years, at the end of which a tensile strength of ~ 1300 MPa was recorded. Young's modulus of the fibres remained virtually unchanged throughout the entire period, suggesting that the structure of the bulk glass is not affected in a major way as a result of reactions with the cement extract at ambient temperatures. The changes in the fibre properties during the first six months were not monitored and it is not certain whether the 1300 MPa represents the strength of the 'man handled' or the reacted fibre.

At 65°C a similar trend (Fig. 1.11(b)) in durability is observed but in this case, as would be expected, the initial reduction in strength was more

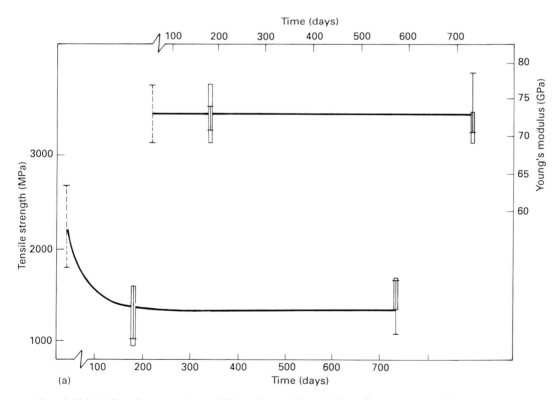

Fig. 1.11(a) Tensile strength and Young's modulus of glass fibres reacted with cement extracts at 20°C. --- pristine fibre, —— fibre from OPC 737 extract; ▢ fibre from OPC 739 extract. The bars include ± one standard deviation.

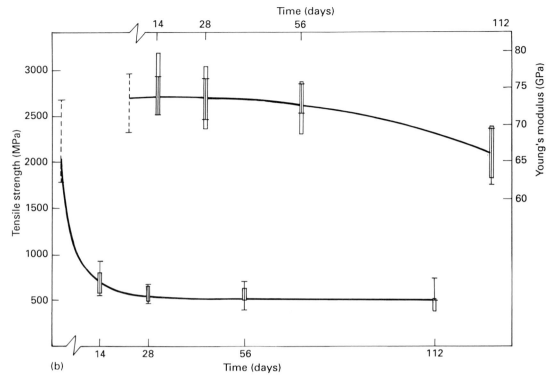

Fig. 1.11(b) Tensile strength and Young's modulus of glass fibres reacted with cement extracts at 65°C. --- pristine fibre, —— fibre from OPC 737 extract; ▢ fibre from OPC 739 extract. The bars include ± one standard deviation.

pronounced and Young's modulus of the fibres also showed a slight reduction after prolonged exposure to the alkaline cement extracts. Again, there is a strong indication that a stable ultimate tensile strength value of 500 MPa was reached after a month.

Oakley and Proctor[43] have reported that the ultimate tensile strength of AR glass fibre strands are reduced from values in the range 1450 to 1750 MPa to 1200 to 1300 MPa after the first 24 h in a cement environment, and at 20°C this strength level is maintained for a few months at least[21]. On the other hand, Cohen and Diamond[44] claimed that when placed in the cement effluent solution of Lawrence[28] at ambient temperatures, AR glass fibre strands made in their laboratory were weakened only slightly over a period of some weeks. These authors speculated that the long-term strength of their fibres could be 80% of their original value.

These results indicate that at ambient temperatures the solution phase in hydrated cements react with AR glass fibres very slowly and the initial physical state of the fibres may be important in determining whether such reactions will produce a reduction in fibre strength so clearly noticeable at slightly elevated temperatures. The physical state of the reinforcement becomes even more important in grc composites. Fortunately the SIC test provides a measure of the fibre strength in the composite that is directly relatable to composite properties. The fibre strength loss with time in the SIC test may be considered to be due to a 'broad corrosion process' which may involve both chemical and physical mechanisms.

Chapter Two
Theoretical Principles

2.1 Introduction

The object of introducing fibres into the cement mix is to improve the mechanical properties and in particular to enable it to withstand tensile loading without catastrophic failure. Without reinforcement the tensile strength of cement is only about 3–5 MPa at best, and cracks soon develop where any tensile stresses occur. If these cracks propagate, failure follows. In compression the strength is very much higher and the ratio of compressive to tensile strength is $\geqslant 10$, a ratio that is typical of 'brittle' materials. Improving the tensile strength will allow the compressive strength to be used to better effect in applications that involve bending where otherwise the low (unreinforced) tensile strength would restrict the bending strength.

The early experience with E-glass reinforcement of high-alumina cement suggested that the incorporation of fibres in the mix leads to a composite that 'yields' in tension rather than fails catastrophically. The result is a material that is more robust in that it withstands impact better than the unreinforced material.

Because of this possibility fibre reinforcement looks a promising method of improving the properties of the otherwise brittle, low-strength material. Before any worthwhile application can be considered, however, it is necessary to understand the way the fibres work. This will allow the sensible design of the composite, and the prediction of the behaviour in more complicated stress situations. It is because of these considerations that so much effort has been applied to developing theories to describe the properties of composite materials and their relationships to the properties of the components, the fibre volume fractions and the way the fibres are arranged.

The earliest workers in this field were Krenchel[1] and Allen[2]. Krenchel developed theories that allow the elastic modulus of the composite to be calculated from the properties, orientation and volume fractions of the constituents. He suggested 'efficiency factors' that allow for the effect of fibre length and orientation on the modulus and strength. Allen produced a model to predict the effect of long fibres on the tensile stress/strain curve, and also showed how the properties in bending can be calculated from the predicted or experimentally measured tensile and compressive stress/strain curves. Aveston, Cooper and Kelly[3] have described the stress/strain curve in detail defining the salient points, and have set the understanding of the mechanism of fibrous reinforcement on a sound footing. The earlier model has been extended[4] to include short fibres and other orientations than that (aligned in the stress direction) of the original Aveston, Cooper and Kelly (ACK) model. Attempts have been made to understand how cracks develop and the conditions that allow them to propagate. In parallel with the theoretical work, considerable effort has been applied to the experimental measurement of the stress/strain properties in tension and in bending and also of the impact resistance. The results have been used to test and to validate the theoretical work.[5]

2.2 Notation

The subscripts c, f and m refer to composite, fibre and matrix; o and ℓ refer to orientation and length, respectively.

The basic symbols are as follows:

E	the elastic modulus
G	the shear modulus
μ	Poisson's ratio
m	($=E_f/E_m$) the modular ratio
σ	stress
σ_u	ultimate or breaking stress
ε	strain
ε_u	ultimate or breaking strain
v	the volume fraction
ℓ	fibre length
a	fibre cross-sectional area
p	fibre perimeter
τ	the interfacial bond strength
τ_s	the static bond strength
τ_d	the sliding or dynamic bond strength
η	the efficiency factor
U	energy to break

Other symbols are defined in the text.

2.3 Mechanism of reinforcement: aligned long fibre composites

The mechanism of reinforcement of the matrix by fibres can be described by first considering the response to stress of a model composite consisting of continuous parallel fibres aligned in the direction of the applied stress, and then describing the modifying effect of fibre orientation and length.

The first effect of the fibre addition is on the elastic modulus: because the modulus of the glass fibres is higher than that of the cement, the initial slope of the stress/strain curve (the modulus) is increased. The stress supported by the composite at a strain ε is then

$$\sigma \, (= E_c \, \varepsilon) = E_f v_f \, \varepsilon + E_m v_m \, \varepsilon \qquad (2.1)$$

and the modulus of the composite follows the simple 'mixture rule':

$$E_c = E_f v_f + E_m v_m \qquad (2.2)$$

where v_f and v_m are the volume fractions of the fibre and matrix, respectively.

The stress at which the matrix cracks is also increased because the modulus is increased, and becomes

$$\sigma_{mu} = E_c \varepsilon_{mu} \qquad (2.3)$$

where ε_{mu} is the matrix failure strain.

In the ACK model of composite behaviour the crack that forms extends completely across the matrix, releasing fibres in that region. If the composite strain is held at the point where the matrix cracks, the stress drops. Provided there are sufficient fibres to support the stress without breaking, on further increase in strain the fibres then extend and slip through the matrix as the load increases, until sufficient stress is transferred to the matrix to form a second crack. The minimum fibre volume needed to allow this to happen, the 'critical fibre volume fraction', $v_{f_{crit}}$ is defined by the inequality

$$\sigma_{fu} v_{f_{crit}} > E_c \varepsilon_{mu} \qquad (2.4)$$

where σ_{fu} is the fibre failure strength.

If the matrix has a well defined single-valued failure strain it will continue to crack in this way, and the stress/strain curve will have a jagged appearance, until the distance between the cracks is too small to allow sufficient stress to be transferred from the fibre to the matrix to break it further. The fibres alone take any further increase in stress, and are pulled through the matrix as they extend, until they break at their failure strength. The composite strength is then that of the fibre contribution alone $\sigma_{fu} v_f$, the matrix affecting the strain at composite failure but not the stress.

The idealised stress/strain curve of the composite, smoothed in the multiple cracking region AB, is shown in Fig. 2.1. The model supposes that load is transferred between fibre and matrix in the multiple cracking region, so that it changes linearly with distance from the crack. If the bond shear strength τ is constant, the distance needed to transfer sufficient stress to break the matrix, i.e. to produce strain in the matrix of ε_{mu} is given by

$$x' = (v_m/v_f)\, \sigma_{mu} a/p\tau \tag{2.5}$$

where $\sigma_{mu} = E_m \varepsilon_{mu}$.

The matrix will crack into blocks of minimum length x' and maximum length $2x'$, after which further load is borne by the fibres alone as they extend and slip through the matrix. The strain at the end of the multiple cracking region, ε_{mc}, is the average strain of the fibres at that time, i.e.

$$\varepsilon_{mu}(1+\alpha/2) < \varepsilon_{mc} < \varepsilon_{mu}(1+3\alpha/4) \tag{2.6}$$

where $\alpha = E_m v_m / E_f v_f$. The upper and lower limits correspond to the minimum spacing x' and maximum spacing $2x'$, respectively.

Since the fibres alone take any further increase in load, the curve beyond this point has slope $E_f v_f$ determined by the fibre modulus and volume fraction. When the local strain in the fibres at the crack reaches the fibre

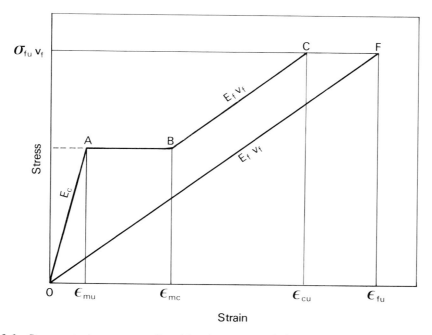

Fig. 2.1 Stress–strain curve predicted by Aveston *et al.* for an aligned continuous (i.e.) long fibre composite.

failures strain the composite fails. It fails at the stress required to break the fibres $\sigma_{fu}v_f$. The composite failure strain is not the failure strain of the fibres, but is reduced because of the load sharing along part of the fibre length. There is in effect a shifting of the stress/strain curve of the fibre alone (OF of Fig. 2.1) as a result of this load sharing, but the slope is unchanged.

The failure strain ε_{cu} is the average strain of the fibres at failure and is given by

$$(\varepsilon_{fu} - \alpha\varepsilon_{mu}/2) < \varepsilon_{cu} < (\varepsilon_{fu} - \alpha\varepsilon_{mu}/4) \tag{2.7}$$

After the fibres break the load falls to zero.

The energy absorbed per unit volume by the composite is the area under the stress/strain curve to failure, and is given by

$$\tfrac{1}{2}\sigma_{fu}v_f\varepsilon_{fu} < U < (\tfrac{1}{2}\sigma_{fu}v_f\varepsilon_{fu} + \tfrac{1}{4}\alpha E_c \varepsilon_{mu}^2)$$

provided, of course, that the volume fibre fraction is equal to or greater than the critical volume fibre fraction, given by equation (2.4).

The energy is determined mainly by the strain energy required to strain the fibres to failure, and the contribution from multiple cracking of the matrix is usually small compared to the elastic strain energy of the fibres.

There is a modification to the above analysis that should be mentioned at this stage, i.e. the crack spacing. Gale, cited in Reference 4, has pointed out a parallel between the actual spacing between cracks and the random parking of cars. Statistically the minimum average space between cars of length x' parked at random in a given space is $(1.364 \pm 0.002)x'$.

The strain at the end of the multiple cracking area then becomes[4]

$$\varepsilon_{mc} = \varepsilon_{mu}(1+0.659\alpha) \tag{2.8}$$

and the composite fails at a strain of

$$\varepsilon_{cu} = \varepsilon_{fu} - 0.341\,\alpha\varepsilon_{mu} \tag{2.9}$$

The energy to break becomes

$$U = \sigma_{fu}v_f\varepsilon_{fu}/2 + 0.159\,\alpha E_c\varepsilon_{mu}^2 \tag{2.10}$$

Kimber and Keer[6] have shown that for composites having large lengths the theoretical value for the average gap between cracks is 1.337. The use of this value in preference to Gale's will result in very minor amendments to equations (2.8), (2.9) and (2.10).

When a rectangular beam of an ideal elastic material is stressed in bending, the neutral axis is in the centre of the beam, and the modulus of rupture (MOR), defined as the stress in the outer surface at failure, is equal to the tensile strength. In common with materials that yield, this does not apply to grc. Once the strain on the outer surface has reached the point at which cracks appear, the stress remains constant (assuming a constant-stress multiple cracking region) at increasing strain, the neutral axis moves towards the compressive face, and the bending moment increases. The apparent modulus of rupture, calculated on the assumption of linear elastic behaviour, increases. In the limit, if the material continued to extend at constant stress, the apparent modulus of rupture approaches a factor of three times the true stress on the tensile face[4].

Tensile stress/strain curves based on the ACK expressions for the strains at the end of multiple cracking and at failure are shown in Fig. 2.2, together with the bending stress/strain curves calculated from them[7]. At the critical fibre volume fraction ($v_{f_{crit}}$) the fibres support the stress after the matrix has cracked in tension, and there is a long multiple cracking region (curve A). Over this region the bending moment continues to rise (curve A') and the ratio of nominal bending strength (MOR) to ultimate tensile strength (UTS) is high. As the fibre volume fraction increases the length of the multiple cracking region decreases, the composite becomes more stiff and the MOR/UTS ratio decreases. At high volume fractions (not shown) the tensile curve approaches that of the fibres alone and the MOR/UTS ratio approaches unity.

Fig. 2.3 shows the relationship between the nominal bending strength and fibre volume fraction calculated from tensile stress/strain curves derived using the ACK model (full lines). The curves apply above the critical fibre volume fraction for reinforcement in tension ($v_{f_{crit}}$). Below $v_{f_{crit}}$ in tension the stress capacity of the fibres after the (single) matrix crack has formed can still lead to reinforcement in bending[8], provided that the fibre volume fraction is above a (lower) critical fibre volume fraction $v'_{f_{crit}}$. At the critical volume fibre fraction $v'_{f_{crit}}$ for reinforcement in bending, the nominal bending strength is equal to the tensile strength of the (reinforced) matrix. The extrapolated response for fibre contents between $v'_{f_{crit}}$ in bending and $v_{f_{crit}}$ in tension[7] is shown by the dotted lines (Fig. 2.3). $v_{f_{crit}}$ for each curve is indicated by its intersection with the (broken) curve A, which marks the change from single to multiple fracture. The relationship between nominal bending strength and fibre volume fraction is non-linear but, over a practical fibre content range, the non-linearity is small. If the best straight line is fitted to data over a limited range there will be a positive intercept on the stress axis and the bending strength/volume fraction relationship might appear to follow a simple 'mixture rule'. This could be very misleading since such an apparent mixture rule for the strength in bending of composites that fail by multiple cracking results from the changing shape of the tensile stress/strain curve as the fibre volume fraction increases. The positive intercept that results from fitting the best straight line to

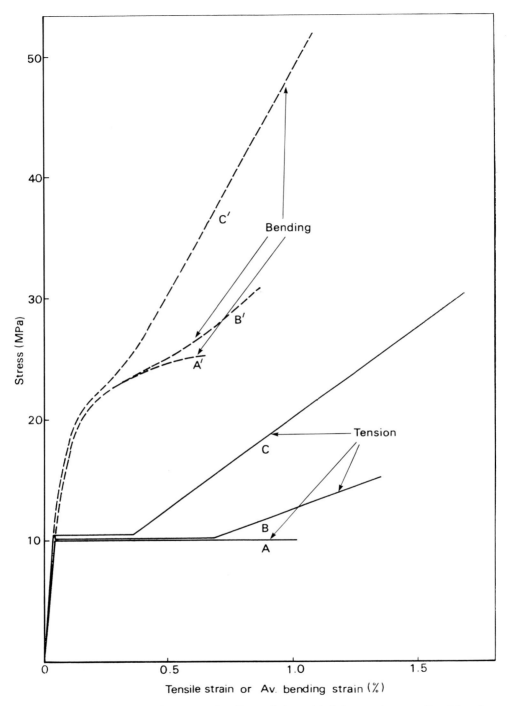

Fig. 2.2 Tensile stress–strain curves for fibre reinforced brittle matrices predicted by the ACK theory (full line), and the bending response calculated from them (broken lines). Aligned long fibre composites, with $E_f = 76$ GPa, $E_m = 25$ GPa, $\varepsilon_{mu} = 0.04\%$ and $\varepsilon_{fu} = 2\%$. Fibre contents are 0.7% (curves A and A′), 1% (curves B and B′) and 2% (curves C and C′).

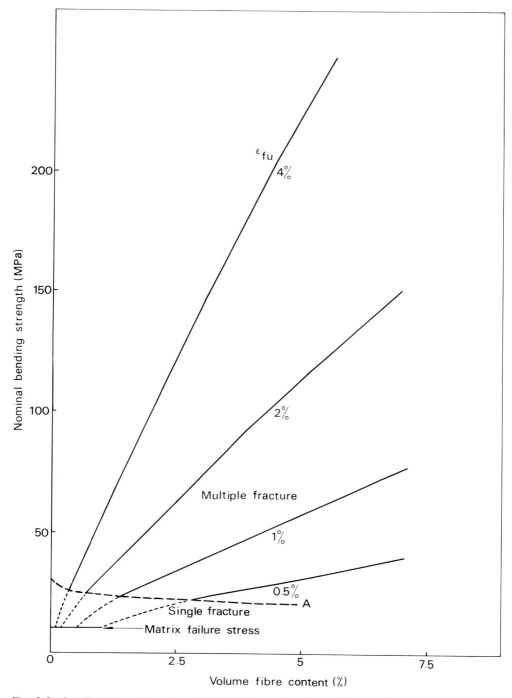

Fig. 2.3 Predicted bending strength (MOR) as a function of fibre volume content for aligned long fibre composites, for the parameters used in Fig. 2.2, except for ε_{fu} values of which are shown. Full lines relate to fibre volume fractions above the critical fibre volume fraction for reinforcement in tension ($v_{f_{crit}}$); dotted lines refer to extrapolated response below $v_{f_{crit}}$.

bending data is not the matrix strength in bending, and any apparent mixture rule for the bending strength is not a true mixture rule.

2.4 Effect of fibre length and orientation: efficiency factors

Although some glass fibre is used in the form of continuous aligned filaments or rovings, the principal glass fibre reinforced cement composites that are in use today are made with short fibres. In sheet material the manufacturing method produces a random planar fibre arrangement; that is the fibres are randomly arranged in planes parallel to the face of the sheet. In pre-mixed grc the fibres are arranged in all three directions, approximately in random three-dimensional (3-D) array. The mechanism of reinforcement generally applies, and in particular stress/strain curves in tension resemble that described for the aligned long fibre model composite. However, there are two important modifications that need to be made; the first involves the effectiveness of fibres that are not aligned with the applied stress; the second concerns the effect of fibre length, since short fibres might slip and pull out rather than reinforce in the way described earlier. These modifications are usually applied to the expressions for modulus and strength, through 'efficiency factors' of orientation and length, and these are outlined below. The effects on energy to break, crack spacing and on the salient points on the stress/strain curve are also discussed.

2.4.1 The elastic modulus

The modulus of glass fibre is approximately 70–76 GPa, and that of cement paste about 20–25 GPa. When glass fibres are combined with cement paste, the result is an improvement in the modulus if the fibres are aligned and continuous or long (equation (2)). In composites where the fibres are not aligned and are short, an efficiency factor η is usually applied to the fibre contribution and this depends on fibre length and orientation to the direction of applied stress. The modulus calculated on this basis may not show an improvement due to fibre addition.

In calculating orientation efficiency factors Krenchel[1] postulated that the fibres can support load only along their length. He calculated the effectiveness of fibres at angles between 90° and 0° to the stress direction. The orientation efficiency factor η_o is a maximum (=1) when the fibres are aligned with the applied stress, and is a minimum (=0) when they are perpendicular to it.

In practice most of the strong grc is produced by the spray-up method, using chopped fibre strands. As mentioned earlier the spray method distributes the fibres in a random way, and the composite sheet is formed by building up successive layers. This leads to a 'random two-dimensional' array of short fibres. An efficiency factor can be calculated to allow for the random distribution of the fibres.

Krenchel calculated the factor for the condition where the composite is constrained at its edges so that it does not contract in the lateral direction. The value of η_o derived for the random two-dimensional array is 3/8. The modulus of a long fibre composite is then

$$E_c = \eta_o E_f v_f + E_m v_m = 3/8 \, E_f v_f + E_m v_m \tag{2.11}$$

In calculating the allowance that should be made for short fibre length it is usually assumed that stress is transferred from the matrix to the fibre linearly along its length from the ends of the fibre until the strain in the fibre reaches that of the matrix. The length of fibre ℓ_x needed to transfer a stress of $E_f \varepsilon_x$, where ε_x is the average strain in the matrix, is given by

$$\tfrac{1}{2}\ell_x = \sigma_x a / p\tau \tag{2.12}$$

where $\sigma_x = E_f \varepsilon_x$ and τ is the strength of the fibre/matrix bond.

Fibres of length ℓ will then have length ℓ_x ($\tfrac{1}{2}\ell_x$ at each end) over which the stress increases from zero to $E_f \varepsilon_x$, and a length $(\ell - \ell_x)$ over which the stress is equal to $E_f \varepsilon_x$.

The average stress supported by the (aligned) fibres is then $E_f \varepsilon_x (1 - \ell_x/2\ell)$, and $(1 - \ell_x/2\ell)$ is the efficiency factor η_ℓ that should be applied to allow for the finite fibre length.

$$\text{From equation (2.12) with } \varepsilon_x = \varepsilon_{fu}, \; \ell_c = 2\sigma_{fu} a / p\tau_s \tag{2.13}$$

where ℓ_c is the critical fibre length, that is the minimum length required for the fibre to reach its breaking strain, ε_{fu}.

The efficiency factor η_ℓ depends on the average strain of the matrix, ε_x, and on the strength of the bond between the fibre and the matrix. η_ℓ can then be written:

$$\eta_\ell = \left(1 - \frac{\ell_c}{2\ell} \frac{\varepsilon_x}{\varepsilon_{fu}}\right)$$

Even at the end of the elastic region, that is when the matrix starts to crack, the strain is very small, i.e. only of the order of a few hundred microstrain for cements and concretes, and the length efficiency factor is unlikely to be less than 0.98 assuming typical values for bond strength, modulus, length, etc.

The total efficiency factor to allow for both length and orientation will be a combination of both orientation and length factors above. Equation (2.2) becomes

$$E_c = \eta E_f v_f + E_m v_m \tag{2.14}$$

where η is a factor that describes the combined effects of orientation and length of fibre in reducing their effectiveness compared with that when the fibres are aligned and 'infinitely' long.

The combined efficiency factor is not simply $\eta_o \cdot \eta_\ell$ since the fibre length required to transfer stress from the matrix, and therefore the contribution to η, depends on orientation. However, the error in using the simple factor $\eta_o \eta_\ell$ is not significant in practice for grc because the fibre volume fraction is small, i.e. only about 5% normally, and the effect of fibres in 'reinforcing' the matrix is less than 2% even for long fibres. The reason for adding fibres is not to improve the elastic modulus but to improve the strength in tension.

2.4.2 Tensile strength

Provided the fibre volume fraction is sufficient (equation (2.4)), the composite undergoes multiple cracking, and as for the aligned continuous fibre composites described earlier, when no further cracks can be formed the fibres alone carry any further increase in stress. The composite strength is then determined by the maximum stress the fibre arrangement across a crack can support. Once the stress on the composite reaches the breaking stress of the weakest fibre arrangement bridging a crack, the other cracks close up as the fibres break and pull-out of the matrix in the weakest region.

The strength of the composite is usually written in the form

$$\sigma_{cu} = \eta_o \, \eta_\ell \, \sigma_f \, v_f \tag{2.15}$$

where η_o is the efficiency factor for orientation and η_ℓ for fibre length, bond strength, etc. In this case η_o and η_ℓ are the strength efficiency factors, and are not necessarily the same as the factors that apply to the elastic modulus.

There is still some controversy about the values of the factors η_o and η_ℓ. Krenchel[1] has used values of η_o previously outlined for the orientation efficiency factor. He used a value of $\frac{3}{8}$ for η_o for a random two-dimensional array, although he has suggested that the value could be higher as the fibres in the cracked region are pulled into alignment. Aveston et al.[4] have used a value of $\frac{1}{2}$. For η_ℓ, the efficiency factor of length, Krenchel has used a value of $\left(1 - \dfrac{\ell_c}{\ell}\right)$. Bortz[9] and Aveston et al.[4], on the other hand have used a length efficiency factor of $\left(1 - \frac{\ell_c}{2\ell}\right)$. Laws[10] has shown that these different factors are limits to a more general expression, and that the two efficiency factors η_o and η_ℓ are not independent.

The analysis of Laws[10] calculates the stress/strain response of an arrangement of fibres across a cracked section. This allows the strength, and therefore the strength efficiency factors, to be derived. It is also used to predict the stress/strain curve of the composite.

It is useful first to outline the way this is done for a composite in which the fibres are short and aligned in the stress direction. Consider fibres of length ℓ aligned along the stress axis and arranged so that the fibre ends are uniformly distributed along the length of the composite. A crack perpendicular to the stress axis is assumed to release the fibres over a crack width (separation of the faces of the crack) which is small compared with the fibre length. The fibres are intersected by the crack so that the shorter lengths of fibre embedded in the matrix are uniformly distributed between 0 and $\ell/2$. At a strain ε_x, those fibres will slip that have an embedded length less than $\ell_x/2$, where $\ell_x/2 = E_f \varepsilon_x a / p \tau_s$, and the probability that a fibre will slip is ℓ_x/ℓ.

The average stress supported by all the fibres is

$$\bar{\sigma}_x = \left(1 - \frac{\ell_x}{\ell}\right) E_f \varepsilon_x = \left(1 - \frac{\ell_c \varepsilon_x}{\ell \varepsilon_{fu}}\right) E_f \varepsilon_x \tag{2.16}$$

where ℓ_c is defined by equation (2.13) above.

From equation (2.16) it follows that the average stress increases with strain to reach a maximum value $(\bar{\sigma}_x)_{max}$ at a strain $(\varepsilon_x)_{max}$, where

$$(\bar{\sigma}_x)_{max} = \tfrac{1}{4}\sigma_{fu}\left(\frac{\ell}{\ell_c}\right) \tag{2.17}$$

and

$$(\varepsilon_x)_{max} = \tfrac{1}{2}\varepsilon_{fu}\left(\frac{\ell}{\ell_c}\right) \tag{2.18}$$

If $(\bar{\sigma}_x)_{max}$ is reached at a value of $(\varepsilon_x)_{max}$ which is less than the fibre breaking strain ε_{fu}, the breaking stress of the mat is defined by this maximum (equation (2.17)). In this case the composite fails by fibre pull-out.

If, however, the fibre breaking strain is reached before this maximum stress is developed, the fibres that have not slipped will break, and this maximum will not be achieved. The strength of the composite will then be given by equation (2.16), where $\varepsilon_x = \varepsilon_{fu}$.

Thus there are two possibilities:

(i) If $\ell < 2\ell_c$,
$$\bar{\sigma}_{fu} = (\bar{\sigma}_x)_{max} = \tfrac{1}{4}\left(\frac{\ell}{\ell_c}\right)\sigma_{fu} \text{ and } \eta_\ell = \tfrac{1}{4}\left(\frac{\ell}{\ell_c}\right) \qquad (2.19)$$

(ii) If $\ell > 2\ell_c$,
$$\bar{\sigma}_{fu} = \left(1 - \frac{\ell_c}{\ell}\right)\sigma_{fu} \text{ and } \eta_\ell = 1 - \left(\frac{\ell_c}{\ell}\right) \qquad (2.20)$$

If, after failure of the (static) interfacial bond τ_s, there is a sliding or ploughing frictional force between fibre and matrix described by a bond strength τ_d, the average stress supported by the fibres that have slipped at a strain ε_x is

$$\bar{\sigma}_d = \frac{\ell_x}{\ell} \tau_d \frac{\bar{\ell}p}{a}$$

where $\bar{\ell}$ is the average length of the fibre ends that have slipped; that is, $\bar{\ell} = \ell_x/4$.

The total average stress supported by the fibres at a strain ε_x is then

$$\bar{\sigma} = \bar{\sigma}_x + \bar{\sigma}_d = \left(1 - \frac{\ell'_c}{\ell}\frac{\varepsilon_x}{\varepsilon_{fu}}\right) E_f \varepsilon_x \qquad (2.21)$$

where $\ell'_c = \tfrac{1}{2}\ell_c(2 - \tau_d/\tau_s)$.

The stress/strain response of the fibre arrangement across the crack, given by equation (2.21), is plotted in Fig. 2.4, for various ℓ/ℓ_c and τ_d/τ_s ratios. These curves replace the linear fibre response OF (Fig. 2.1) in the derivation of the stress/strain curve of the aligned short fibre composite. This is described later. The curves are also used to calculate the strength of the composite and to derive the strength efficiency factors. They show, for example, that the maximum stress occurs at a strain below fibre failure strain if the critical fibre length equals the fibre length, and $\tau_d/\tau_s < 1$ (curves A and B). If $\tau_d = 0$, the maximum stress is at $\tfrac{1}{2}\varepsilon_{fu}$ when $\ell = \ell_c$ (curve A); and a fibre length twice the critical fibre length is needed before the failure strain of the arrangement reaches ε_{fu} (curve D).

The efficiency factors in equations (2.19) and (2.20) are modified accordingly:

(i) If $\ell < 2\ell'_c$ $\quad \eta_\ell = \dfrac{\ell}{2\ell_c(2 - \tau_d/\tau_s)} = \tfrac{1}{4}\left(\dfrac{\ell}{\ell'_c}\right) \qquad (2.22)$

(ii) If $\ell > 2\ell'_c$ $\quad \eta_\ell = 1 - \dfrac{\ell_c}{2\ell}(2 - \tau_d/\tau_s) = 1 - \left(\dfrac{\ell'_c}{\ell}\right) \qquad (2.23)$

The length efficiency factor for fibres satisfying the condition $\ell > 2\ell'_c$ varies between $(1 - \ell_c/2\ell)$ and $(1 - \ell_c/\ell)$ depending on the ratio τ_d/τ_s. The bond

Theoretical Principles 39

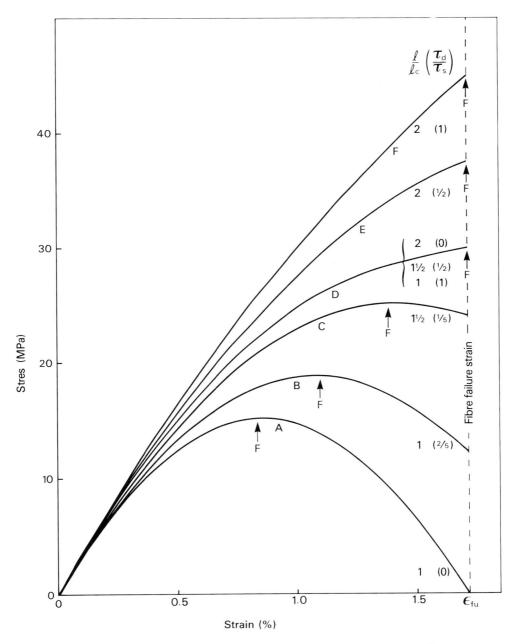

Fig. 2.4 Stress plotted against fibre strain for various ratios ℓ/ℓ_c and τ_d/τ_s, for an arrangement of aligned, short fibres held across a crack. $E_f = 70$ GPa, $\sigma_{fu} = 1200$ MPa and $v_f = 0.05$. 'Failure' occurs at F.

strength measurements of de Vekey and Majumdar[11] suggest that τ_d/τ_s for cements could be as large as $\tfrac{2}{3}$, leading to an efficiency factor η_ℓ of $(1 - 2\ell_c/3\ell)$.

When the fibres are all parallel and the composite is stressed in a direction making an angle θ with the direction of alignment, the analysis follows that above, but the strain in the fibres is not necessarily equal to that applying in the simple aligned case. Aveston et al.[4] assume that the fibres are aligned in the stress direction where they cross the crack, but there are fewer of them and the analysis is simply modified by a factor allowing for this. Laws[10] assumes that the crack is very narrow compared with the fibre diameter. In this case, the strain ε_f of the fibre along its axis is related to the strain in the stress direction ε_x by

$$\varepsilon_\theta = \varepsilon_x \cos^2\theta \tag{2.24}$$

Equation (2.16) then becomes

$$\sigma_\theta = \left(1 - \frac{\ell_\theta}{\ell}\right) E_f \varepsilon_\theta = \left(1 - \frac{\ell_c}{\ell} \frac{\varepsilon_\theta}{\varepsilon_{fu}}\right) E_f \varepsilon_\theta \tag{2.25}$$

where $\ell_\theta = 2 E_f \varepsilon_\theta\, a/p\tau_s$.

If there are n fibres per unit length uniformly separated, the number of fibres crossing unit length of crack face is $n \cos\theta$. The average stress in the x-direction is then

$$\bar{\sigma}_x = \bar{\sigma}_\theta \cos^2\theta \tag{2.26}$$

and

$$\bar{\sigma}_x = \left(1 - \frac{\ell_c}{\ell} \frac{\varepsilon_x \cos^2\theta}{\varepsilon_{fu}}\right) E_f\, \varepsilon_x \cos^4\theta \tag{2.27}$$

The fibres break when the (maximum) strain along their length is equal to the breaking strain ε_{fu}, i.e. when

$$\varepsilon_\theta = \varepsilon_x \cos^2\theta = \varepsilon_{fu} \tag{2.28}$$

Hence

$$\varepsilon_x = \varepsilon_{fu}/\cos^2\theta \tag{2.29}$$

which is greater than ε_{fu} when $\theta > 0$.

For a composite that fails by fibre breakage the efficiency factors are then $\eta_\theta = (1 - \ell_c/\ell)$ and $\eta_\theta = \cos^2\theta$ assuming no sliding frictional forces operate ($\tau_d = 0$).

The stress/strain response and the strength of other fibre arrangements are obtained by summation. For a random two-dimensional arrangement it is assumed that there are fibres arranged in all directions, and the average stress in the x-direction follows by integration from equation (2.25), whence

$$\sigma_x = \tfrac{3}{8}\left(1 - \frac{5}{6}\frac{\ell'_c}{\ell}\frac{\varepsilon_x}{\varepsilon_{fu}}\right)E_f\varepsilon_x \qquad (2.30)$$

The full analysis is given in Laws[10] and Laws et al.[12, 13]. The analysis allows for a dynamic or sliding frictional force once the fibres have begun to slip, and leads to stress/strain responses similar in shape to those shown for the aligned case (Fig. 2.4). These curves apply to the (fictitious) fibrous mat, and are used later to derive the stress/strain curve of the composite. To do this the load-sharing between fibres and matrix between successive cracks is taken into account.

The maximum loads/stresses at 'failure' are those at the crack, i.e. those calculated for the fibre mat above, and those that apply in the composite. The efficiency factors are given in Table 2.1.

Table 2.1 Efficiency factors for strength, restrained fibrous mat.

Orientation	Continuous fibres	Efficiency factor: short fibres	
Aligned	1	$\tfrac{1}{4}\ell/\ell'_c$	$(\ell \leq 2\ell'_c)$
		$1 - \ell'_c/\ell$	$(\ell \geq 2\ell'_c)$
Random two-dimensional	$\tfrac{3}{8}$	$\tfrac{9}{80}\ell/\ell'_c$	$\left(\ell \leq \tfrac{5}{3}\ell'_c\right)$
		$\tfrac{3}{8}\left(1 - \tfrac{5}{6}\tfrac{\ell'_c}{\ell}\right)$	$\left(\ell \geq \tfrac{5}{3}\ell'_c\right)$
Random three-dimensional	$\tfrac{1}{5}$	$\tfrac{7}{100}\ell/\ell'_c$	$\left(\ell \leq \tfrac{10}{7}\ell'_c\right)$
		$\tfrac{1}{5}\left(1 - \tfrac{5}{7}\tfrac{\ell'_c}{\ell}\right)$	$\left(\ell \geq \tfrac{10}{7}\ell'_c\right)$
		where $\ell'_c = \tfrac{1}{2}\ell_c(2-\tau_d/\tau_s)$	

There are small modifications that strictly should be made to the above relationship depending on the nature of static and dynamic frictional bond strengths. These are outlined in Hannant[14]. It has been pointed out earlier that there are other proposed strength efficiency factors of length and of orientation, and it has proved difficult to confirm experimentally which of these is correct.

It is convenient, however, to express the total efficiency factor for strength of a random 2–D composite that fails by fibre fracture in the form

$$\eta = \eta_o \eta_\ell = \eta_o \left(1 - c \frac{\ell_c}{2\ell}(2-\tau_d/\tau_s)\right) \qquad (2.31)$$

where c is a factor having a value of $\frac{5}{6}$ or 1.

2.4.3 Stress/strain curves in tension and in bending

Earlier in this chapter, the idealised stress/strain curve for an aligned continuous fibre composite was described. The curve is based on expressions derived for the crack spacing, and for the strains at the end of the multiple cracking zone and at failure. The stresses corresponding to these strains were assumed to be the stress at first matrix crack, $E_c \varepsilon_{mu}$ and the stress $\sigma_{fu} v_f$ supported by the fibres when they fail at strength σ_{fu}. The 'curve' was then constructed by connecting these salient stress/strain points.

These curves are used to derive an analytical expression for the strength in bending, and for the ratio MOR/UTS of strength in bending to strength in tension.

When the fibres are short and/or not aligned, the stress/strain curve is less simply defined, and an analytical expression for the strength in bending and for the MOR/UTS ratio probably cannot be derived.

An important difference between the stress/strain curve for the continuous aligned fibre composite described by Fig. 2.1, and that of a short fibre composite, is the shape of the curve beyond the end of the multiple cracking zone. For continuous aligned fibres, the response in this region is linear, reflecting the normal stress/strain curve of the fibres that bridge the matrix cracks (OF of Fig. 2.1). Laws[15] has suggested that the linear response of the continuous fibres should be replaced by the non-linear response of the arrangement of short fibres that bridge the cracks. The non-linear response arises because the proportion of fibres that slip increases as the crack widens. This is illustrated in Fig. 2.4, for a composite in which the short fibres are aligned in the direction of the applied stress. When the fibres are short and are not aligned, similar non-linear curves result[13].

The other information required in order to calculate the stress/strain curve is the effect of orientation and/or fibre length on the crack spacing.

2.4.3.1 CRACK SPACING, STRAINS ε_{MC} AND ε_{CU}, AND STRESS/STRAIN CURVE IN TENSION

Aveston *et al.*[4] have suggested that, for their model composite consisting of long fibres randomly arranged in two dimensions, the strength of the composite σ_{cu} is $\frac{1}{2} \sigma_{fu} v_f$ if the fibres break, and the crack spacing is $\frac{\pi}{2}$ times that of the corresponding aligned composite.

For short fibres, their expression for the minimum crack spacing is more complicated[4], namely

$$x_d = \tfrac{1}{2} [\ell \pm (\ell^2 - 4\ell\, x_2)^{\frac{1}{2}}] \tag{2.32}$$

where x_2 is the minimum crack spacing for continuous fibre composite (i.e. $x_2 = x'$ for aligned and $x_2 = \tfrac{\pi}{2} x'$ for random 2−D composites). x_d is the length needed for short, randomly oriented fibres to transfer sufficient load ($\sigma_{mu} v_m$) to the matrix to break it.

The strains ε_{mc} and ε_{cu} of the composite depend on the distribution of (average) stress in the fibres as a function of distance from a crack. For aligned long fibres it is usually assumed that stress transfer is linear, and the relationships previously given result. In other cases, the stress distribution is not known and no detailed analysis has been made as far as the authors know.

Proctor[16] has proposed factors K_o and K_s to allow for orientation and length in the expressions for ε_{mc} and ε_{cu}, so that

$$\varepsilon_{mu}(1 + \alpha/2K_oK_s) \leqq \varepsilon_{mc} \leqq \varepsilon_{mu}(1 + 3\alpha/4K_oK_s) \tag{2.33}$$

and

$$K_s \varepsilon_{fu} - \alpha\varepsilon_{mu}/2K_oK_s \leqq \varepsilon_{cu} \leqq K_s\varepsilon_{fu} - \alpha\,\varepsilon_{mu}/4K_oK_s \tag{2.34}$$

Laws[15] has queried these expressions, because the efficiency of reinforcement of short fibres depends on the stress, and the factor K_s at the end of matrix cracking would be different from the factor K_s at fibre failure.

Assuming that stress is transferred to the matrix linearly from the crack and the crack spacing is twice the transfer length (for $\sigma_{mu} v_m$ to be transferred), the expression derived by Laws for the strain at the end of multiple cracking is

$$\bar{\varepsilon}_{mc} = \varepsilon_{mu}\left(1 + \frac{K_{mu}}{K_{mc}}\right)/2 + \alpha\, \varepsilon_{mu}/2\eta_o K_{mc} \tag{2.35}$$

where η_o is the orientation efficiency factor, and K_{mu} and K_{mc} are length efficiency factors that correspond to the fibre strain $\bar{\varepsilon}_{mu}$ at the centre of the block and ε_{mc} at the end.

Usually $K_{mc} \sim K_{mu}$ so that the expression for ε_{mc} reduces to

$$\bar{\varepsilon}_{mc} = \varepsilon_{mu} + \tfrac{1}{2}\alpha\, \varepsilon_{mu}/\eta_o K_{mc} \tag{2.36}$$

which, since $K_{mc} \sim 1$, reduces to

$$\bar{\varepsilon}_{mc} = \varepsilon_{mu}(1 + \tfrac{1}{2}\alpha/\eta_o) \tag{2.37}$$

The expression for $\bar{\varepsilon}_{cu}$ when fibres fail becomes

$$\bar{\varepsilon}_{cu} = \tfrac{1}{2}(1 + \eta_\ell/K_{cu})\,\varepsilon_{fu} - \tfrac{1}{2}\alpha\, \varepsilon_{mu}/\eta_o K_{cu} \tag{2.38}$$

44 *Glass Fibre Reinforced Cement*

where η_ℓ is the length efficiency factor of strength and K_{cu} the length efficiency factor appropriate to the strain of the fibres in the centre of the block.

The predicted stress/strain curves for random 2–D composites are shown in Fig. 2.5 for decreasing fibre length (curves A to D). With decreasing

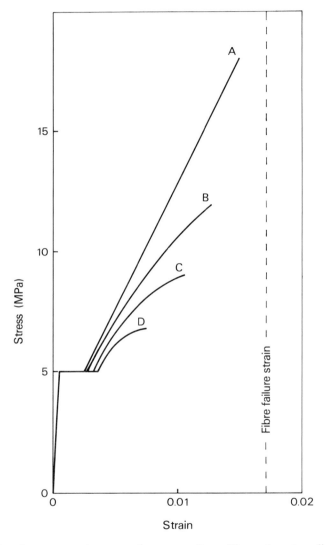

Fig. 2.5 Predicted stress–strain curves for composites with random two-dimensional arrangement of fibres. Curve A is for long continuous fibres. Curves B, C and D refer to decreasing fibre lengths. $E_f = 70$ GPa, $E_m = 20$ GPa, $\sigma_{fu} = 1200$ MPa, $\sigma_{mu} = 5$ MPa and $v_f = 0.04$.

length there is a small increase in the strain at the end of the multiple cracking zone while the failure strain is decreased significantly. The slope of the curve in the post-cracking region of the short-fibre composite is initially close to that of the continuous fibre composite (curve A), and decreases as the strain increases.

After the stress/strain curve for the short fibre composite has been predicted, the bending curve can be calculated from it. To date this has not been done, although bending curves have been predicted[2,17,18] from experimentally measured tensile curves. The MOR/UTS ratio can then be calculated.

2.4.4 Energy considerations

In the preceding section, the energy consumed in the complete fracture of an aligned continuous fibre composite was defined as the area under the stress/strain curve to failure. Expressed as the energy per unit volume, this is given by Hannant, Hughes and Kelly[19] as (see equation (2.10))

$$U = \tfrac{1}{2} \sigma_{fu} \varepsilon_{fu} v_f + 0.159 \, \alpha E_c \, \varepsilon_{mu}^2$$

where the crack spacing has been taken as that derived statistically by Gale (see Reference 4).

In principle a similar analysis can be made for a composite where the fibres are short, and also when the fibres are distributed for example in a random two-dimensional arrangement, as they are in grc. Difficulties arise, however, because there is no universally accepted theory predicting the detailed stress/strain curve for these composites. An estimate can be made assuming correction factors such as those made by Proctor[16], or by Aveston et al.[4] for the salient points on the stress/strain curve, and assuming a linear post-cracking region (after Aveston et al.[4]). The theory outlined by Laws[15] is the most comprehensive available. It makes no assumptions about the values that should be attributed to the efficiency factors of orientation and length but assumes that the stress transfer from fibre to matrix is linear with distance from a crack. Assuming that this is a sufficiently good approximation, the theory can be used to calculate the energy consumed in stressing the composite to 'failure'. It is not feasible to produce analytical expressions for the energy, and a numerical approach is needed.

The energy required to extend and to break the composite can be calculated approximately from the stress/strain curve of the fibre mat alone, neglecting the constraint of the matrix in reducing the average fibre strain at failure. This allows the effects of various parameters on the work to be examined.

It can be shown [20] that for $\ell > \ell_c$ the contribution W of the fibres to the composite strain energy per unit area up to fibre failure strain is

$$W = \frac{3}{16}\left\{\left(1 - \frac{5\ell_c}{9\ell}\right) + \frac{5\ell_c\tau_d}{18\ell\tau_s}\right\}\sigma_f\varepsilon_f v_f L \qquad (2.39)$$

where the first term is the elastic contribution and the second is that arising from frictional slip. L is the length of the specimen. The relative size of the two terms follows from equation (2.39); for example, if $\tau_d = \tau_s$, the ratio of frictional to elastic energy is $\frac{5}{26}$ when $\ell = 2\ell_c$ and increases to $\frac{5}{8}$ for $\ell = \ell_c$.

For $\ell > \ell_c$, the total energy per unit area, W', needed to pull out the fibres that do not break, is

$$W' = \frac{2}{\pi}\left(\frac{1}{12}\frac{\tau_d\ell_c^2}{\tau_s\ell}\sigma_f v_f\right) = \frac{1}{6\pi}\frac{\tau_d r^2}{\ell}\left(\frac{\sigma_f}{\tau_s}\right)^3 v_f \qquad (2.40)$$

For a model composite containing 5% (by volume) of fibres of strength 1200 MPa, length 30 mm and critical fibre length 15 mm, the energy to break calculated from equation (2.40) is approximately 24 kJ m^{-2} when $\tau_d = \tau_s$. The strain energy to maximum stress calculated from equation (2.39) is 0.17 kJ m^{-2} per millimetre of sample strained, or 10 kJ m^{-2} if a specimen of length 50 mm is uniformly strained. Of this, less than 2 kJ m^{-2} is the frictional component.

The contribution of the fibres to the total energy needed to break a specimen and pull out the fibres is (approximately) the sum of the irrecoverable part of the strain energy W and the pull-out energy W'. If the elastic strain energy is largely recovered as one crack opens and the others close, the total energy to break for the model above is approximately 25 kJ m^{-2}, which is within the range of measured impact strengths of young glass-reinforced cement samples.

There has been some discussion of the relative importance of the bond strength and the fibre tensile strength in determining the energy needed to break a fibrous composite. The ratio σ_f/τ_s is directly related to the critical fibre length ℓ_c and determines the proportions of fibres that slip and that hold. From equation (2.39) it follows that a high elastic energy contribution from the fibres requires a low ℓ_c together with a high fibre strength. It therefore requires a strong bond. The maximum possible value of this elastic energy is $\frac{3}{16}\sigma_{fu}\varepsilon_f v_f L$. The maximum work to pull-out, however, requires that all the fibres slip and that τ_d is as high as possible, i.e. that $\tau_d = \tau_s$ since $\tau_d \leq \tau_s$. The relation between pull-out energy, fibre strength σ_{fu} and bond strengths τ_s and τ_d is illustrated in Fig. 2.6. The discontinuities in the curves mark the conditions where all fibres slip; further increase in σ_{fu} then has no effect on pull-out energy, but an improvement in τ_d has a marked effect.

There has been some debate on the appropriate way to describe the

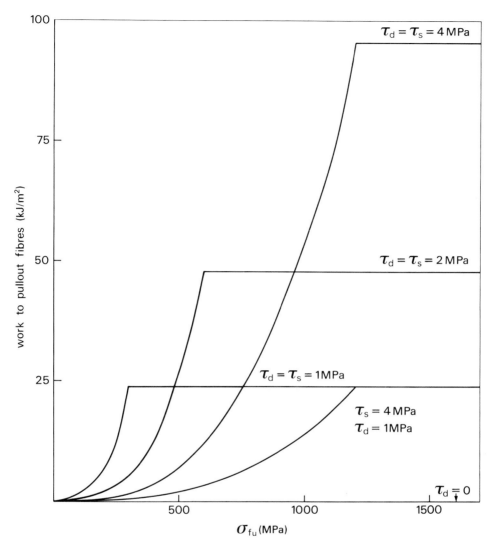

Fig. 2.6 Work to pull-out fibres, calculated from equation (2.40), as a function of fibre strength σ_{fu} and bond strengths τ_d and τ_s (for $v_f = 0.05$, $\ell = 30$ mm and fibre radius $= 0.1$ mm).

energy-absorbing capacity of fibrous composites. The energy per unit volume up to the maximum stress might be considered a minimum or critical energy needed before one crack opens up and catastrophic failure (at constant stress) results. The energy available for fibre pull-out beyond the maximum stress is related to the length over which the stress is applied, and is therefore not a useful composite parameter.

2.4.5 Stress and strain at first matrix crack

The strength of the composite at the point where the matrix cracks is given by equation (2.3):

$$\sigma_{mu} = E_c \varepsilon_{mu}$$

where ε_{mu} is the matrix failure strain, and E_c the composite modulus, given by equation (2.14):

$$E_c = \eta E_f v_f + E_m v_m$$

where η is the combined efficiency factor at the matrix failure strain.

The matrix strength depends on the sum of the stresses supported by the two components, and equation (2.3) is to that extent a 'mixture' rule. However, the stress carried by the fibres is not determined by the fibre strength, since the strain is the failure strain of the matrix and not that of the fibre. The failure strain for a cement matrix is of the order of 0.04%.

The matrix strength, then, depends on the composite modulus, and is not in theory significantly increased by the fibres since the volume fraction is usually only about 5% and the modular ratio is fairly low (about 4, say). For a 5% volume fraction aligned continuous fibre composite the expected improvement would be about 15% and if the fibres were randomly distributed in a plane the improvement would be only $\sim 2\frac{1}{2}\%$.

In the early 1960s Romualdi and Batson[21] suggested that the cracking strain of concrete could be increased by the addition of fine fibres. Romualdi and Batson were concerned with the effects of incorporating fine wires in concrete; nevertheless the idea is interesting to those concerned with the development of glass reinforced cement since improvement in the failure strain and, by implication, the failure stress would lead to higher allowable design stresses.

Experimental evidence was found, however, that the addition of glass fibre leads to an improvement in both stress and strain at first cracking of the cement matrix. It is difficult to establish how much of the observed increase is due to other factors because, for example, the inclusion of fibres in the matrix leads to a porosity change in the matrix. Despite these problems there does appear to be a real effect, although it is very much less than that predicted by Romualdi and Batson.

Aveston et al.[3] produced an apparently different theory to explain the observed increase in strain at first crack. Their theory, which was based on the energy before and after the crack was formed, led to the following expression for the cracking strain in aligned continuous fibre composites:

$$\varepsilon_{mu} = \left[\frac{12\tau\gamma_m E_f v_f^2}{E_c E_m^2 r v_m} \right]^{\frac{1}{3}} \tag{2.41}$$

where γ_m is the surface work of fracture of the matrix and r is the fibre radius.

The ACK model predicted that the matrix cracking strain is increased provided that ε_{mu} calculated from equation (2.41) is greater than the (unreinforced) matrix strain at failure. There is thus a cut-off or discontinuity in the strain/fibre volume fraction curve.

In a subsequent analysis Hannant et al.[19] pointed out that the ACK formula gives a lower limit to the strain that must be exceeded for cracking to occur. In agreement with Romualdi and Batson[21] they suggested that the fibres act by imposing a 'closing force' across a crack, which reduces its opening. Hannant et al. predicted a continuous increasing curve relating first matrix cracking strain to fibre volume fraction, provided that the fibre spacing is less than the critical flaw size according to Griffith's equation.

Hannant et al.[19] also considered the crack size in relation to the inter-fibre spacing, and so determined the limit of the fibre volume fraction for a given fibre size and configuration. When the conditions were such that the fibres intersected the crack they calculated the stress developed in those fibres. For grc assuming typical values for the modulus, etc. and a bond strength τ of 3 MPa, the stress in the fibre when the matrix cracks is 290 MPa, which is very much greater than the fibre stress at the failure strain of the matrix (28 MPa).

2.5 Fibre/cement bond

2.5.1 The effect of fibre/cement bond on composite properties

After the first crack is formed across a fibre cement composite, the matrix stress is thrown on to the fibres at the crack. If the fibres are discontinuous, whether they can support the stress or pull out from the matrix depends on the shear stresses developed at the fibre/matrix interface and on the strength of the interfacial bond.

The theories of fibre reinforcement outlined earlier assume that the shear stress developed at the interface is uniform along the fibre length. The load is then transferred from the fibre to the matrix and decreases linearly with distance from the crack. This assumption is obviously not strictly correct since cement is brittle rather than ductile and does not show a yield stress. Furthermore, it leads to the result that pull-out load is directly proportional to embedded fibre length, and experimental results, particularly for steel fibre/cement specimens, show that this is not so. In the case of a brittle resin matrix, Kelly[22] reported that the resin and fibre become debonded – the debonding could be seen readily within a transparent resin. Kelly also showed that after debonding the frictional stress developed between fibre and matrix is roughly constant during pull-out. He concluded that provided the embedded length is not too long, the assumption of uniform shear stress was sufficiently good.

The effect of elastic stress transfer on the pull-out of fibres from a matrix has been considered by Greszcuk[23] and later by Lawrence[24]. Lawrence showed that in pulling out a fibre from an elastic matrix the shear stress developed was not uniform and the load transferred between the fibre and matrix did not change uniformly with length along the fibre.

The analyses of Greszcuk and Lawrence, mentioned above, show that the shear stress developed when a load is applied to the fibre is a maximum at the point where the fibre enters the matrix. This is illustrated in Fig. 2.7 taken from Lawrence's analysis[24]. When a load is applied such that the shear stress at this point reaches the shear strength of the interface, the fibre debonds completely and pulls out with no further increase in load.

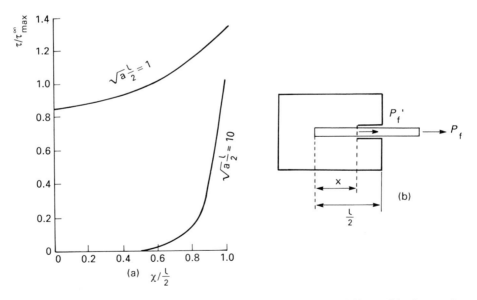

Fig. 2.7 (a) Variation of shear stress with length along embedded fibre. τ^α is the maximum stress developed for an infinitely long fibre; (b) Debonded fibre configuration.

Bond strength calculations which assume a constant shear stress along the embedded fibre length are therefore likely to be low even if the embedded fibre length is short.

Lawrence also considered what would happen if there were a frictional stress τ_i opposing pull-out after the interfacial bond (strength τ_s) had failed. He showed that if the embedded fibre length were greater than a critical value x_{max} debonding would stop and a further increase in load would be needed to continue debonding and lead to complete pull-out. Fig. 2.8 shows the variation in maximum fibre load with embedded fibre length for various ratios of τ_s/τ_i. The curves indicate that for a sufficiently short embedded fibre length the fibre load varies approximately linearly with length along the fibre. Fig. 2.8 shows also that as the frictional bond strength τ_i approaches the interfacial bond strength τ_s, the relationship between pull-out load and fibre length tends to become linear.

The effects of embedded fibre length and elastic properties on the shear stress and load distribution along the fibre length are marked. Lawrence

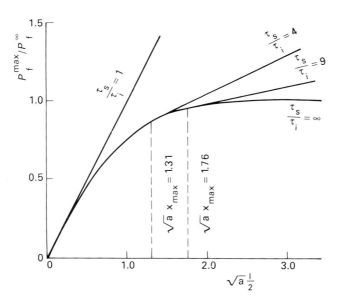

Fig. 2.8 Variation of maximum fibre load with embedded fibre length factor for various friction conditions.

showed that for a given value of α, a parameter based on the elastic properties and geometry of the pull-out test, the difference in these distributions for long and short fibres can be highly significant. The maximum fibre load necessary to cause complete debonding and eventual pull-out is dependent on the length of the embedded fibre and the ratio between the shear strength and the frictional 'shear strength' of the fibre–matrix interface. Further, the development of debonding of the fibre from the matrix can have a marked effect on the maximum shear stress developed at the

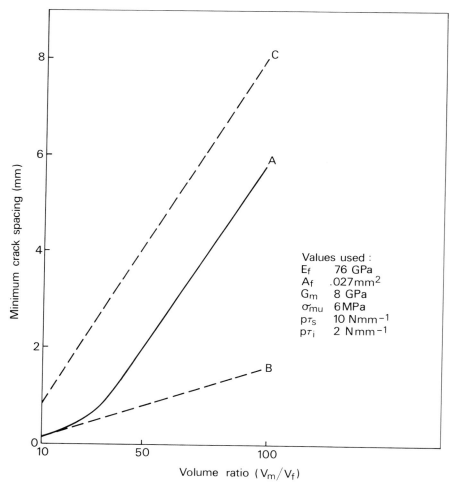

Fig. 2.9 Minimum crack spacing as a function of volume ratio V_m/V_f. Curve A assumes elastic stress transfer and $\tau_s \neq \tau_i$ (crack spacing $\frac{1}{2}\ell_{1,2}$); curves B and C assume linear stress transfer with bond strengths τ_s and τ_i, respectively (spacings $x(\tau_s)$ and $x(\tau_i)$, respectively).

interface. Consequently, if frictional forces play any part in the pull-out mechanism, it is essential to differentiate clearly between catastrophic and non-catastrophic debonding in order to determine the true shear strength of the interface.

Lawrence described how the bond strengths τ_s and τ_i could be estimated from the maximum load/embedded fibre length curve. Bartos[25] later pres-

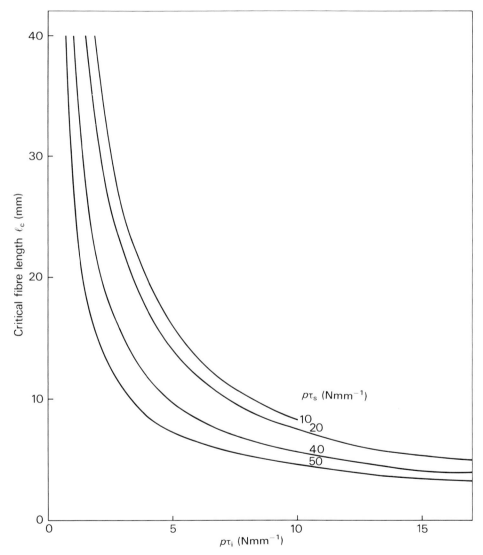

Fig. 2.10 Critical fibre length as a function of $p\tau_i$ for various values of $p\tau_s$.

ented an analysis similar to that of Lawrence to show how the bond strengths can, in some cases, be determined from pull-out curves at a single fibre length.

The implications of the Lawrence study on composite properties has been considered by Laws[26]. If the interfacial bond strength τ_s is equal to the frictional bond strength τ_i, the 'elastic' analysis reduces to the simpler plastic-matrix theory, and the critical fibre length, crack spacing, composite strength, etc. are those outlined earlier. Otherwise the crack spacings (Fig. 2.9) lie between the values obtained using the plastic-matrix theory and assuming constant τ_s (curve C) and τ_i (curve B). The variation of critical fibre length with τ_s and τ_i is shown in Fig. 2.10.

The elastic stress transfer theory was earlier applied by Laws *et al.*[12] to the calculation of the strength of a random two-dimensional short fibre composite. The expression for strength is complex and its evaluation involves numerical integration. The results suggest, at least for composites such as grc, that an increase in the frictional shear stress τ_i has a much greater effect than a corresponding increase in the interfacial bond τ_s. This is not surprising when the maximum fibre pull-out load as a function of embedded fibre length is considered (Fig. 2.8).

Chapter Three
Production Methods for Grc Components

Since its birth in the early 1970s the grc industry has grown at a fairly rapid rate[1]. New applications of grc have emerged regularly and to meet these demands suitable production techniques have been developed. These techniques and the properties and uses of the material and components made of it have been described in detail in the proceedings of various conferences, particularly those organised by the GRC Association. A recent account of state-of-the-art production methods for grc components is given in the monograph by True[2]. An earlier monograph by Young[3], a report by Smith[4] and the 'Cem-FIL grc Technical Data' prepared by Pilkington Brothers PLC[5] for their licensees are other important sources of information on this subject. In this chapter a brief summary only of grc production methods is given, based in part on the work at BRE by Ryder[6] and Hills[7] in the early years of development of the grc industry.

The two most widely used production methods for grc which were tried out at BRE in the late 1960s are the 'premix' and the 'spray' processes. In 'premix' methods the constituents are mixed together to produce a paste. This is subsequently formed into components using well-known techniques such as casting and press-moulding. The 'spray' processes are derived from the grp industry. In these a slurry of cement and sand together with bundles of glass fibres chopped *in situ* are deposited on to a suitable mould. Other grp production methods such as filament winding or lay-up techniques are also applicable in grc manufacture but so far these have been used on a very limited scale. For large-volume production of grc sheets it has been possible to adapt standard asbestos cement manufacturing methods such as the Hatschek or Magnani process.

3.1 Constituent materials

The essential materials in the manufacture of grc are glass fibres, cement, sand and water. Additions, such as pfa or silica fume are sometimes made, and generally speaking the use of admixtures has increased steadily. It is common nowadays to use dispersion and workability aids such as methyl cellulose or air-entraining agents such as lignosulphonates and, to achieve appropriate production schedules, retarders and accelerators are often found to be useful. The addition of a small amount of a suitable polymer dispersion such as acrylics is now recommended. If the grc component is required to have a special colour for aesthetic reasons pigments can be added to the slurry either in powder or liquid form.

The glass fibres used are most commonly the alkali-resistant varieties described in Chapter 1. 'Rovings' or 'chopped strands' of these fibres are now commercially available in many countries. A typical sample of roving contains about 40 strands or ends with approximately 200 filaments in each strand. The individual filaments are $10-20$ μm in diameter. Rovings normally separate into strands when they are chopped, but by changing the coating on the filaments the extent to which the strands break down into filaments can be varied considerably. Soft coatings allow the strands to disperse into filaments when they are mixed with water, whereas harder coatings are more resistant to this treatment and the strands tend to remain as integral units. The extent to which the rovings and strands break down during manufacture is an important factor in influencing the strength of the grc product. It also affects the production process itself.

3.2 Spray production methods

In the spray method the chopped fibre and the cement slurry are sprayed, usually from separate nozzles, on to a mould where they mix as they impinge on the surface. The cement or sand/mortar slurry is poured through a coarse sieve to remove any lumps and fed to the spray gun through a metering pump unit where it is atomised by compressed air. The water:cement ratio must be high enough to achieve a workable mix, yet sufficiently low to avoid undue effects on strength. In order to achieve this, admixtures are often used. Where high water:cement ratios are used to provide the required workability the excess water must be extracted after spraying. The glass fibre in the form of roving is fed to a chopper/feeder mounted on the spray head. This chops the fibre strands to desired lengths and propels them into the mortar stream and gradually a uniform felt of fibre and mortar is built up on the moulds. Many grc products made by the spray-up process contain a nominal 5 wt% of glass fibres.

3.2.1 Manual spray methods

In the manual spray method the operator moves the spray-head backwards and forwards across the mould, directing the stream of material as far as possible perpendicular to the surface, until the required thickness has been built up. Thickness control is achieved by using pin gauges. When the required thickness has been reached, the material is subjected to roller-compaction to ensure that it is free from entrapped air and that the slurry has fully impregnated the fibre. Roller compaction is also important in ensuring compliance with the mould face and in increasing composite density. The rolled surface may finally be trowelled smooth or decorated by sprinkling a suitable aggregate on to the surface and pressing it in by hand tamping. One surface of the sheet produced by this process has an ex-mould finish and the other a rolled, trowelled or decorated finish. After the spraying, the products are covered with polythene sheet, normally demoulded the following day and then transferred to a curing chamber.

As a commercial process manual spraying is labour intensive, the output of a single hand unit being typically 10–12 kg of sprayed product per minute[5]. Its advantage is that quite complex shapes can be produced. The method has been used for the manufacture of profiled cladding panels, facade elements, formwork, ducting, street furniture, agricultural components and so on. It is very versatile and can, for example, be carried out as a site process, to produce grc renderings.

3.2.2 Mechanised spray method

The process described above has been mechanised to produce components that are substantially flat or of shallow profile. In the mechanised method developed by BRE, shown in Fig. 3.1, the mould remains stationary and the spray head traverses it in a programmed fashion. In some commercial designs the mould is moved by means of a roller or flat conveyor and passes under a reciprocating spray. Forward and transverse speeds are balanced and feed sprays are controlled to ensure uniform deposition and correct fibre distribution in the composite. The finished product has an ex-mould finish on one side and trowelled, rolled or decorated finishes on the other similar to those given by the manual method. Outputs can be substantially higher than those achieved by manual methods, and are normally up to 25–30 kg/min.[5]

Various types of grc components, e.g. cladding panels and facade elements, formwork and sewer linings, are now produced by the mechanical spray

Fig. 3.1 Mechanical spray method developed at Building Research Establishment.

process. Shaped components such as ducts and channels are produced by using moulds that fold. Automatic profile-spray machines have been developed recently[8].

3.2.3 Spray-dewatering process

The spray-dewatering process was originally developed and used at BRE as an improved manual (spray) method. It was later mechanised (Fig. 3.1) and has since been adapted for full scale mechanised production of sheet materials.

To allow water to be extracted the horizontal mould on which the sheet is formed must have a perforated base and be covered by a high wet strength paper or porous plastic material which acts as a filter. After a grc layer of suitable thickness has been built up it is dewatered by suction from below. By this means a slurry with a water/cement ratio, typically of 0.55, can be densified to a water/cement ratio of 0.25–0.30 with a correspondingly increased initial strength. The upper surface of the sheet can be given a trowelled or decorated finish if required while the mould side has a mat filter paper texture.

Asahi Glass has developed a machine for the continuous production of grc sheets by spraying and dewatering[9]. The line speed of this machine is about 6–12 m/min, and this rate compares well with that of asbestos cement manufactured by the Hatschek process.

3.2.4 Folding and forming

After the sheet has been sprayed and, where appropriate, dewatered, it may be allowed to set in the flat state and be pressed if necessary. Alternatively it may be formed whilst still in the 'green' state to produce articles of various shapes and designs. The forming of complex shapes is usually carried out manually, but simpler designs such as standard corrugations may be mechanised. Vacuum forming of complex shapes in grc has been achieved by using male or female moulds.

3.2.5 Miscellaneous spray processes

3.2.5.1 PIPES

Short lengths of pipe were produced at BRE by the equipment shown in Fig. 3.2. The mould is rotated at 100 rev/min while spraying the premix grc slurry on to the inside surface. This technique produces a random array of fibres on concentric cylindrical surfaces. A given thickness of fibre-reinforced material can be sprayed to form the outer layer of the pipe. The interior can then be made up to the required thickness by adding unreinforced material, such as a conventional concrete mix. A 1:2 sand:cement mix can also be sprayed, adding glass fibre reinforcement to the outer layer as before. After the spraying, the excess water is spun off, usually by increasing the speed of rotation of the mould. The pipe sections can be demoulded after curing for about 12 hours.

3.2.5.2 RENDERS AND SCREEDS

At BRE the spraying technique was used to apply a grc rendering to an aerated concrete block wall. In such applications the wall can be built without mortar joints, by stacking the blocks on top of each other and spraying with 60% cement and 40% pulverised fuel ash slurry containing 4% glass fibre. If the blocks are fairly dry, the surplus water in the render is removed by absorption by them.

The spraying technique can also be used on non-porous surfaces, where it

Fig. 3.2 Experimental spinning equipment for grc pipes, one of which is shown.

tends to produce a material of a lower density that might be useful either for thermal insulation or fire protection. These renderings are easily trowelled to give a fair-faced surface as the fibre addition allows the surface to be worked but holds the bulk of the material in place.

GRC formulations containing 0.5–2.5% by weight Cem-FIL AR glass fibre have been developed for renders and screeds[5].

Spraying processes produce grc of good strength and toughness as the fibres are distributed in the matrix in random 2–D orientations. These reinforce the matrix more efficiently than premix processes (described later), where the fibres are distributed randomly in all directions. Consequently, the spray-up methods of grc manufacture, both manual and automatic, have been very popular in the grc industry. There have been recent developments in essential equipment such as the spray pump unit and the spray head. Of particular interest has been a new design which combines the mortar spray head with the fibre chopper gun by locating the former around the nozzle of the latter. This development, known as the concentric spray head, produces a spray pattern that considerably reduces wastage of materials. Many of these improvements have been described by True[2].

3.3 Mix and place methods

Methods which involve mixing chopped fibre strands with cement, sand, water and admixtures before casting are referred to as 'premix' methods. The process consists of two stages, namely the production of a slurry that has the right workability to allow the uniform mixing of the fibre, and the blending of the fibres into the slurry. The fibres are added after the slurry has been mixed so that fibre damage is kept to a minimum.

The early work at BRE showed that tangling and matting of the glass fibre in the mixer lead to poor fibre dispersion in the matrix and to difficulty in handling and placing the mix. The presence of fibres provides an easy route for water to escape from the body of the mix to the surface, and hence segregation and water loss occur even at very low applied pressures. A standard technique adopted at BRE[7] to overcome some of these difficulties was to modify the wet mix properties by the addition of very small amounts of polymers such as polyethylene oxide and methyl cellulose. By dispersing the fibres in the polymer solution before adding the solids and mixing, a degree of lubrication of the fibres is achieved and the aqueous phase viscosity is increased. This produces better fibre distribution in the mix and helps water retention.

Premixed grc was produced at BRE experimentally using a number of standard and modified proprietary mixers. With low speed pan mixers, some fibres were found to collect on the leading edges of the blade and pan side scraper. Despite these problems, a pan type mixer was preferred to a drum mixer, where balling or matting of fibres posed greater difficulties. Current practice in the industry is to blend the matrix by high shear mixing, then to transfer it to a paddle type mixer and have the fibres sprinkled in quickly to avoid balling. The mixing time is kept to a minimum commensurate with good dispersion of the fibre and adequate 'wet-out' of the composite. Premix grc plants can be fully automated.

Mix formulations used in premixed grc depend on the products being made, sand contents of 50% by weight of cement being fairly common. Workability at low water:cement ratio, preferably not exceeding 0.35, is often achieved by using admixtures. Up to 4% by weight of Cem-FIL fibre, 12–25 mm long, have been used in premixed grc products[5].

3.3.1 Processing methods

Several processes used extensively in the precast concrete industry are suitable for shaping the wet mix of cement mortar and glass fibre.

Gravity moulding in open or semi-open moulds has been used to produce street furniture such as litter bins and planters, junction boxes, electrical transformer housings, etc. Fairly complex shapes can be formed by this technique using moulds made of steel, timber or grp. For intricate designs, for instance in sunscreens or screen walling panels, polystyrene or rubber moulds are sometimes preferred. Compared to normal concrete practice much thinner sections can be cast in grc and the mix is usually less dense and of reduced fluidity. External vibration is commonly applied to remove air voids and to assist the flow of the mix.

Injection moulding techniques have been used to produce window frames and hollow columns. Premix grc containing up to 5% by weight of glass fibre and admixtures has been processed by this technique using Mono pumps, peristaltic pumps and high output concrete pumps. Pumping action causes extra fibre damage, and removal of air is difficult to achieve.

For relatively simple flat objects such as tiles or lids the wet grc premix can be pressed and dewatered using equipment that may vary in complexity depending on the output desired. For high volume production fully automatic systems have been designed. At BRE three types of press were used in the trials, namely a paving flag press on a mould dewatered on both faces, an experimental press on a mould with vacuum dewatering on one face and a low pressure press dewatered on both faces.

Since much of the water is expelled during pressing the initial water content of the mix can be as high as is necessary for handling and mould filling. An admixture is frequently used to retain water during the initial period of pressing and to assist the fibre mix to flow and fill the mould uniformly before consolidation and dewatering. The amount of admixture used is strictly controlled depending on pressing conditions such as pressure and speed of pressing.

A variation of the mix and place method has been developed in which the 'premix' is produced by spray-up and then transferred to the mould in a suitable way. Fibre damage is reduced considerably in this method.

3.4 Other (miscellaneous) processes

3.4.1 *The winding process*

This method is widely used in the fibre reinforced plastics industry, particularly for making components in the high technology area. In 1973 the Concrete Society Technical Report[10] on fibre reinforced cement composites stated that the winding process 'promises to become the most elegant method of producing composites of exact properties'. In the grc industry only a limited use of this method has been made so far, perhaps due to cost considerations. The method is illustrated schematically in Fig. 3.3.

Glass fibre rovings are first impregnated with cement or mortar paste by passing them, sometimes in an opened-up state, through a well-stirred bath of cement slurry. They are then wound round a suitable rotating former or mandrel and at the same time additional slurry and chopped fibre strands are sprayed on the framework. Excess slurry and surplus water are then removed by pressure or suction to produce a well-compacted grc composite. It is customary to keep the fibre strands under slight tension so that they remain straight and aligned.

The method is eminently suitable for making pipes, fence posts, etc., and sheets are produced easily. In the 'green state' the sheet can be shaped to give a variety of products. Grc with high fibre content, 15–20% by volume, can be made by the winding process, and the distribution of the fibre in the composite is usually very uniform. The continuous and aligned nature of the fibre in the product provides a highly efficient form of reinforcement.

3.4.2 The lay-up process

This method is also adapted from the well-established procedures in the fibre reinforced plastics industry. For the production of grc composites various forms of the glass fibre reinforcements — roving, chopped fibre mats or fabric — can be used and objects of various sizes and shapes can be made by employing suitable moulds or forming surfaces. Efficient penetration of the reinforcement by the cement or mortar paste is difficult to attain, and it is advantageous to 'soak' the reinforcement in cement slurry beforehand. After the reinforcements are placed in the mould or on the forming surface, further amounts of cement paste are applied with special rollers and in suitable cases the mix consolidated with pressure. Compaction by vibration is also possible and water surplus to the requirement is removed by suction or pressure.

3.4.3 Asbestos cement methods

After a considerable amount of industrial research and development work it has now become possible to use glass fibres in place of asbestos in the traditional asbestos cement production methods, e.g. the Hatschek or Magnani process. Products such as sheets or pipe manufactured by these processes are cheaper than those made by other methods described earlier. Ryder[6], among others, has given an account of the various asbestos cement

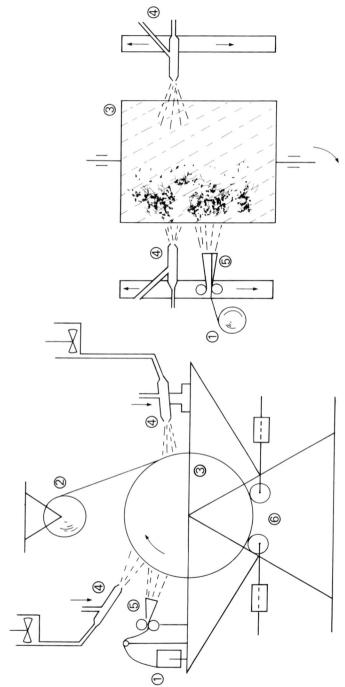

Fig. 3.3 The fabrication of glass fibre reinforced cement (grc) by the winding process (Reference 10): (1) bobbin with roving or strand; (2) roving or strand; (3) winding cylinder; (4) cement paste sprayer; (5) chopped fibre sprayer; (6) compression rollers.

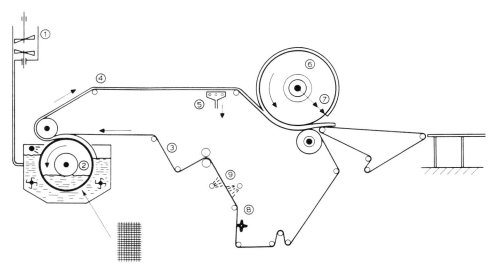

Fig. 3.4 The fabrication of asbestos-cement sheets by the Hatschek process (Reference 10): (1) mixer agitator; (2) screen cylinder; (3) felt band; (4) ply of asbestos cement; (5) dewatering; (6) calender; (7) cutter; (8) beater; (9) sprayer.

production techniques, and Smith[11] has described the different types of asbestos cement products that have been made in grc.

The essential features of the most widely used Hatschek method are shown schematically in Fig. 3.4. The process begins with the preparation of a dilute suspension of very short fibres and cement (typically 6% solids when asbestos is used). The suspension is agitated continuously (for about 2 h in the case of asbestos), when the suspended solids are picked up as a thin film on the surface of a rotating cylinder of fine wire mesh. It is transferred to an endless conveyer belt of permeable felt and passed over a vacuum box to remove excess water. The film is then transferred to a steel assimilation drum on which it is further dewatered and compacted by a pressure roller and is plied to the desired thickness. When the required thickness has been built up on the calender, the deposit is cut with a knife and it peels off on to another endless conveyor belt from which this 'wet flat' is transferred to a platform for further processing. The 'wet flat' is moulded, by hand or mechanically, to form flat or profiled sheets or more complex shapes. Finally, the products are taken to the curing room.

Several problems were encountered in the early attempts to replace asbestos by glass fibres in the Hatschek process. These included non-uniform distribution of the fibre in the product, depletion of cement from the film and hence high solids contents in the recycling water and low water retention in the layers. The addition of cellulose and modification of the glass fibre surface have helped to overcome some of these problems, and with other improvements that are being attempted at present it is hoped that the new generation of asbestos cement replacement products made from grc will meet the requirements of various national and international standards.

3.5 Curing

In common with most products made from hydraulic cements grc needs to be cured in a carefully controlled manner. At any given time the properties of cements reflect their degree of hydration, and this in turn depends on the method of curing. Since in most grc products cement plus sand (and/or pozzolana) constitutes 95% or more of the total solids, proper curing of grc is a very important consideration. When an appropriate curing schedule is strictly followed, the cement matrix in grc hydrates uniformly throughout, ensuring good reproducibility of specified properties in the product. Many grc products are thin and it is essential that they are cured in a moist environment to minimise early shrinkage as far as possible. Only properly cured grc will have enough strength, stiffness and toughness to be handled safely after demoulding.

The details of a curing shedule for grc depend on several factors such as the type of cement used and the mix design, the product and the manufacturing method employed. According to Pilkington's Technical Data manual[5] a satisfactory curing regime for grc should consist of (1) covering the product immediately after manufacture with a polythene sheet until demoulding after 16–20 h and (2) storage in a 'fog room' kept at 15–20°C and greater than 95% RH. Total immersion under water is sometimes used for the second stage. It is also good practice to allow the product, following moist curing, to come to some sort of equilibrium with the ambient conditions in a gradual manner.

It is possible to increase the rate of strength development in grc by increasing the curing temperature or by using chemical accelerators. These procedures need to be controlled very carefully, as pointed out by True[2], if consistent and acceptable strength results are to be attained. In this respect a recent development has been the use of some polymer dispersions, e.g. acrylics, in mix formulations. The addition of the polymer dispersion reduces the long, standard curing period of grc products[12].

3.6 Surface finishes

Surface finishes are important in many applications of grc on aesthetic as well as durability grounds, i.e. in cladding panels and outdoor facades or internal ceilings and wall surrounds. Many of the surface finishes used on grc products have been adapted from the concrete industry. True[2] has discussed the advantages and disadvantages of most of these techniques. Surface finishes on grc need to be compatible with the highly alkaline cement-rich mix, and their effect on moisture movements in the product requires careful consideration. Non-uniform moisture movements from the two faces may contribute to bowing.

Some types of finish, e.g. an exposed aggregate finish, are applied to grc components prior to demoulding and curing and hence may be viewed as part of the overall production process. Such a finish can be obtained by sprinkling or placing fairly small aggregates on to the uncured surface of the grc product and tamping. Alternatively, a layer of aggregate or a mixture of aggregate, sand and cement can be laid on the face of the mould before spraying up the grc material. If necessary the aggregates can be exposed after the product has been cured by acid etching or sand blasting. By using an etch retarder on the mould surface the glossy finish on grc, which is prone to crazing, can be transformed to a sand textured finish. Care should be taken to prevent exposure of the glass fibre on the etched surface by choosing an appropriate thickness for the mist coat.

An epoxy resin finish to the surface of a grc component can be obtained by gel moulding. In this process, a special two-part epoxy resin system is first placed on the mould surface and the grc material sprayed immediately afterwards. It is also possible to bond thin plastic films of, say PVC, to the surface of a grc component in this way.

There are finishes that are applied on to the grc surface after the product has been demoulded and/or cured. Some of these have been described by True[2]. Since these techniques cannot strictly be considered to be part of the grc production process they are not discussed here.

3.7 Quality control

The quality of a fibre composite material such as grc where the proportion of fibre is rather small, 5 wt % or less, depends very strongly on the amount and fibre parameters such as its length and orientation. The quality also depends on the uniformity of fibre distribution in the matrix and how well the latter has been compacted around the fibre to provide an intimate bond. During the production of grc care is taken to estimate some of these quality controlling parameters by continuously monitoring the fibre consumption from the weight loss in the roving and the mortar pump output. Very often a test piece is made which is similar in all operational respects to the production. The fibre content of the composite is determined routinely from samples taken out from this test piece by the wash-out method. Product thicknesses are checked by pin or indentation test gauges or by other methods. The slump characteristics of the slurry and the water/solids ratio of uncured grc are also used as process controls. For quality assurance of the product, wet and dry bulk density (of the material), water absorption and apparent porosity are measured using standard techniques. Mechanical properties such as the limit of proportionality and modulus of rupture are also determined. The quality control aspects of grc production have been discussed more fully by Ward and Proctor[13] and by True[2].

Two of the important parameters that have an enormous influence on the properties of grc made by the spray or premix methods are fibre orientation and distribution. It is difficult to obtain quantitative information on this subject but a technique first studied by Hibbert[14] has been partially successful. Hibbert has shown that glass fibre strands in a thick ($\sim$ 7 mm) section of grc composites are capable of transmitting quantities of light that can be recorded with simple equipment even when the incident light intensities are relatively modest (Fig. 3.5). Using such a technique Hibbert and Grimer[15] were able to establish a correlation between the variations in the number of visible strands in several grc samples and their fatigue properties. In this work a contact print or a photograph of the dark face of the specimen recorded the numbers of the light-transmitting elements. Later work by

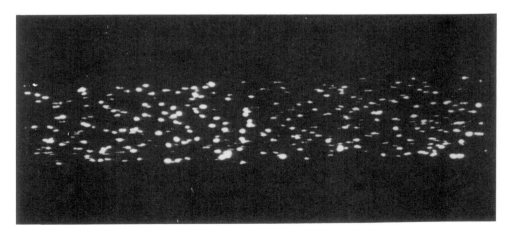

Fig. 3.5 Contact print of a grc section.

Rayment and Majumdar[16] has demonstrated that a scanning technique using a photodiode can be used successfully for the same purpose. They pointed out that for cement samples the curing history is important as the degree of interaction between the cement and the glass has a marked influence on the glass transmittance.

Ashley[17] has described a method of estimating the glass content in grc composites containing Cem-FIL type AR glass fibres based on the determination of zirconium by X-ray fluorescence.

Quality control in the grc industry is being increasingly exercised by the use of National Standards or recommendations issued by national or international bodies. For instance, BS 6432:1984 stipulates the test methods to be used for determining the properties of the grc material, and the Prestressed Concrete Institute in the USA have recently published their report 'Recommended Practice for GFRC Concrete Panels.' The GRC Association has produced several reports to help their member firms control the quality of their products. Among these, 'Specification for alkali resistant glass fibre rovings and chopped strands for reinforcement of cements and concretes' (S 0105), and 'Specification for GRC : GRC 21' (S 0108) are particularly relevant. A degree of quality control is also embodied in the granting of certificates to specific grc products by various national organisations such as the Agrément Board in the UK.

Chapter Four
Properties of Portland Cement Grc

Portland cements are by far the most important hydraulic cements used in the building industry. In grc manufacture both ordinary and rapid hardening varieties are widely used and a quantity of fine silica sand, about half the weight of cement, is commonly included in the cement slurry to reduce drying shrinkage in the composite. For particular purposes, e.g. marine use, special types of Portland cements such as sulphate-resistant Portland cement are recommended. The constituents of the matrix are already covered by British Standards, e.g. for cement and sand, but the appropriate form of a product standard for alkali-resistant glass fibres has yet to be developed.

Although the use of premixed grc has grown in recent years, the utilisation of the reinforcing ability of glass fibres is less in this product than in sprayed-up grc. Much of the property data that are available in the current literature were obtained with the latter form of grc and specifically with spray-dewatered grc having approximately 5 wt% of the original Cem-FIL alkali-resistant glass fibres, 34–38 mm long and a water/solids ratio of 0.28–0.35. Most of the properties described in this chapter relate to such a composite.

4.1 Spray-dewatered grc

4.1.1 Mechanical properties

The mechanical properties of sprayed-up grc depend upon mix formulations, in particular on the length and volume fraction of the fibre used and on curing. These properties change with time, to different extents in different environments, reflecting the changes in matrix and fibre properties and at the interface. Much research has been done in the past decade to determine the magnitude of these changes and to develop the methodology for predicting the very long-term properties of grc.

4.1.1.1 NEAT OPC/GRC

The properties of spray-dewatered grc made from ordinary Portland cement and 5 wt% of 34–38 mm long Cem-FIL AR glass fibres and kept in three different environments up to 20 years are given in Table 4.1. For such a composite the 28 day values of compressive strength and Poisson's ratio are 60–100 MPa and 0.24–0.25, respectively, and the strain to failure in direct

Table 4.1 Measured mean strength properties of spray-dewatered opc/grc at various ages (5 wt% Cem-FIL AR glass fibre).

Property	Total range for air and water storage conditions at 28 days	1 year			5 years		
		Air*	Water+	Weathering	Air*	Water+	Weathering
(a) Bending							
MOR (MPa)	35–50	35–40	22–25	30–36	30–35	21–25	21–23
LOP[1] (MPa)	14–17	9–13	16–19	14–17	10–12	16–19	15–18
(b) Tensile							
UTS (MPa)	14–17	14–16	9–12	11–14	13–15	9–12	7–8
BOP[2] (MPa)	9–10	7–8	9–11	9–10	7–8	7–9	7–8
Young's modulus (GPa)	20–25	20–25	28–34	20–25	20–25	28–34	25–32
(c) Impact strength (Izod) (kJ/m^2)	17–31	18–25	8–10	13–16	18–21	4–6	4–7

Property	10 years			17 years	20 years	
	Air*	Water+	Weathering	Weathering	Air*	Water+
(a) Bending						
MOR (MPa)	31–39	17–18	15–19	13–18	30–32	13–15
LOP[1] (MPa)	14–16	16–17	13–16	11–14	10–14	13–14
(b) Tensile						
UTS (MPa)	11–15	6–8	7–8	4–7	10–14	—
BOP[2] (MPa)	9–10	6–8	—	—	~5	—
Young's modulus (GPa)	25–33	25–31	27–30	25–36	24	—
(c) Impact strength (Izod) (kJ/m^2)	15–22	2–3	2–6	2–5	17–22	~2

* At 40% relative humidity and 20°C
+ At 18–20°C
[1] Limit of proportionality
[2] Bend-over point marking the onset of gross matrix cracking

tension is usually of the order of 1%. Measured in-plane and interlaminar shear strength values lie in the range 10–17 and 3–5 MPa[1].

The 17 and 20 year strength results given in Table 4.1 refer to a rather limited number of samples that have been tested recently. Coupons cut from 1 m × 1 m grc panels exposed to weather at Garston in both vertical and horizontal positions over 17 years have given better results. For instance, the MOR of these coupons was 19–20 MPa, LOP 13–16 MPa, UTS 6–7 MPa, impact strength (IS) 4–5 kJ/m^2 and Young's modulus 31–35 GPa/m^2. The ultimate tensile failure strain of these coupons was in the range 250–300 microstrain compared to 150–170 microstrain obtained for the small coupons exposed to weather directly (Table 4.1).

Majumdar and his colleagues have investigated the effect on fibre length and fibre content on the properties of spray-dewatered grc kept in different environments for up to ten years[2–4]. Their 28 day bending, tensile and impact strength results are shown in Figs. 4.1–4.3. The effect of fibre volume fraction and fibre length on the tensile stress-strain diagrams of grc composites is illustrated in Figs. 4.4 and 4.5.

It is evident from Figs. 4.1–4.3 that for all fibre lengths used, the bending, tensile and impact strength of grc increase with increasing fibre proportions up to about 6% by volume. But whereas the impact strength of the composite continues to increase beyond 6 vol % fibre addition, both MOR and UTS show a reduction. The most likely explanation of this

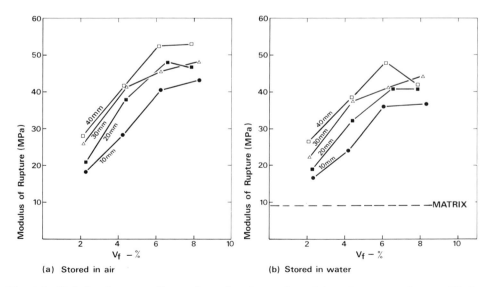

Fig. 4.1 Relation between fibre volume fraction and modulus of rupture of grc at 28 days for different fibre lengths: (a) stored in air; (b) stored in water.

Properties of Portland Cement Grc 73

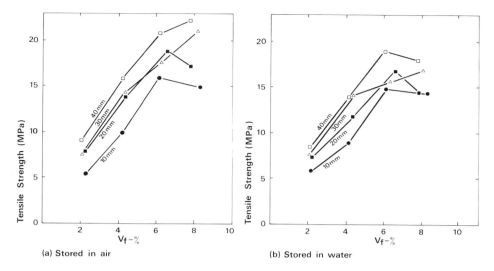

Fig. 4.2 Relation between fibre volume fraction and tensile strength of grc at 28 days for different fibre lengths: (a) stored in air; (b) stored in water.

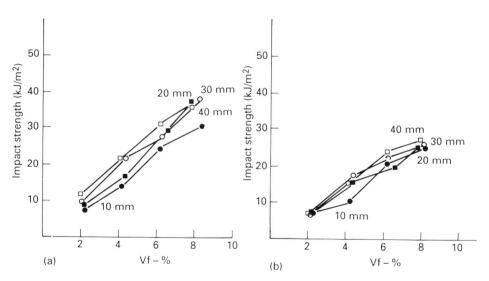

Fig. 4.3 Relation between fibre volume fraction and impact strength of grc at 28 days: (a) stored in air; (b) stored in water.

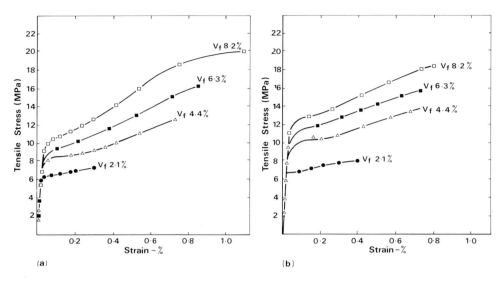

Fig. 4.4 Tensile stress–strain curves of grc composites containing 30 mm long fibres with different fibre volume fractions at 28 days: (a) stored in air; (b) stored in water.

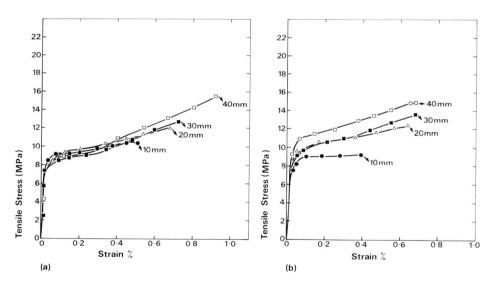

Fig. 4.5 Tensile stress–strain curves of grc composites containing 4 vol% fibres with different fibre lengths at 28 days: (a) stored in air; (b) stored in water.

Properties of Portland Cement Grc

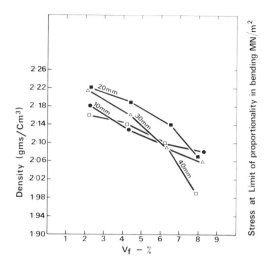

Fig. 4.6 Relation between fibre volume fraction and density of grc composites at 28 days for different fibre lengths.

difference is the rapid decrease in the density of grc that follows the incorporation of fibres beyond a certain amount (Fig. 4.6). As the fibre volume increases, the distribution of the fibre in the composite becomes less uniform, and the penetration of the fibre bundle by the matrix and the compaction of the composite become more difficult. It has also to be borne in mind that glass fibre strands, which are the reinforcing elements in grc, comprising, say, 200 individual filaments, have a built-in porosity and its contribution to the total porosity of the composite is obviously dependent on the fibre content. An increase in the porosity of grc due to excess fibre addition impairs the development of bending and tensile strength of the composite by reducing the total fibre/cement contact area, but this is an advantage as far as the impact resistance of the material is concerned.

Figs. 4.1–4.3 also show that at 28 days the strength properties of grc increase with increasing fibre length, although the effect is less pronounced than that of increasing fibre content. It can also be seen that in common with bending and tensile strength, but perhaps to a greater degree, the impact strength of grc is higher in air than in water. When cured in water, cement hydrates to a greater extent than in air, the products of hydration filling the voids in the matrix as well as in the fibre strands. The overall porosity of the composite is thus reduced relative to air storage, leading to some loss in pseudoductility. Lower bending and tensile strength in water storage is probably associated with some reduction in fibre tensile strength brought about by the corrosive action of the cement matrix.

4.1.1.2 EFFECT OF AGE ON GRC STRENGTH

In grc both major components, i.e. the cementitious matrix and the glass fibres, show significant changes with time. Some of the properties of the composite, e.g. compressive strength, elastic modulus, Poisson's ratio, LOP, BOP, interlaminar shear strength, etc., are controlled predominantly by the matrix whilst others, notably the bending and tensile strength, depend mainly on the properties of the fibre. The changes in the matrix-controlled properties of grc with age reflect the degree of cement hydration and hence are different for different environments as exemplified by Young's modulus results given in Table 4.1.

The strength properties of grc that are mainly controlled by the fibre show considerable changes with age in wet environments. In such environments the tensile strength of the glass fibre reinforcement is gradually reduced with time due to the corrosive action of the cement matrix (see Fig. 4.12) and the strength of the composite decreases correspondingly. The data given in Table 4.1 show that while over a period of ten years in dry air the bending, tensile and impact strengths of grc made from OPC and Cem-FIL AR glass fibres remain substantially unaltered, under continuously wet conditions and in natural weathering these properties fall significantly. The 17 and 20 year results are also somewhat lower than those at ten years. Typical changes in the tensile and bending properties of grc (made from original Cem-FIL fibre) over five years are shown in Figs. 4.7 and 4.8. It is clear that in a wet and natural weathering environment the excellent initial pseudoductility of the material is lost with age and the composite becomes essentially brittle. There is some indication that the reduction in properties may be slower in large components.

The strength properties of grc with varying glass contents and glass lengths have been determined at five years[3] and ten years[4] and a summary of the ten-year results is given in Table 4.2. These results indicate that the long-term strength retention by the composite in natural weather can be slightly increased by increasing fibre volume fractions, and in dry air the composites remain pseudoductile for at least 10 years (Fig. 4.9). There are no significant improvements when fibre lengths are increased. The matrix cracking stress and strain values at ten years show an increase with an increase in fibre volume (Fig. 4.10).

4.1.1.3 GRC/(OPC + SAND)

The trends in the durability of grc composites containing varying proportions of sand stored in relatively dry air and completely wet environments listed in Table 4.3 follow a similar pattern to that observed in the case of grc made from neat OPC[5,6]. From this table it is clear that the bending strength (MOR) of the composites with or without sand addition remains essentially

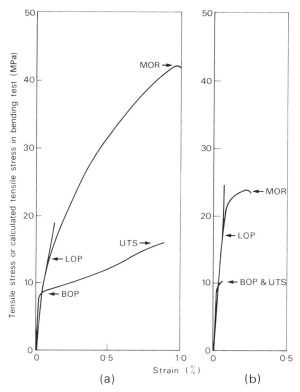

Fig. 4.7 Representative stress–strain curves in tension and bending: (a) after 28 days in water at 18°C to 20°C; (b) after 5 years in water at 18°C to 20°C.

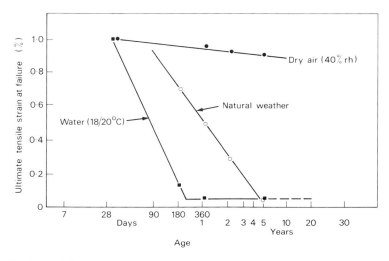

Fig. 4.8 Strain-to-failure in tension for grc at various ages in dry air, natural weather and water storage.

Glass Fibre Reinforced Cement

Table 4.2 Properties of grc composites at ten years for varying glass contents and glass lengths.

Glass		Properties							
		MOR (MPa)			UTS (MPa)			IS (kJ/m^2)		
Content (vol %)	Length (mm)	Air	Water	Weather	Air	Water	Weather	Air	Water	Weather
2.1	10	15.2	13.3	10.6	5.3	3.6	3.6	5.8	2.8	5.0
	20	15.7	14.5	11.0	4.9	4.5	3.0	7.6	2.9	4.3
	30	18.5	18.6	10.6	5.6	3.9	3.1	8.8	5.5	3.1
	40	19.4	14.7	15.0	6.5	5.6	4.6	8.9	2.0	3.2
4.4	10	23.8	18.3	17.2	7.8	5.2	5.3	13.3	2.9	6.2
	20	30.6	20.3	20.4	11.3	7.4	7.1	17.4	4.1	6.5
	30	36.0	19.4	22.4	12.8	7.9	7.7	20.7	3.9	5.2
	40	36.6	21.2	21.7	12.9	7.4	8.0	19.0	3.0	5.7
6.3	10	35.2	24.5	23.0	12.9	10.3	8.4	19.3	5.0	7.0
	20	41.5	24.7	27.4	15.1	10.9	9.9	25.6	3.9	11.5
	30	39.5	24.6	25.0	14.0	11.8	8.7	24.0	4.6	8.6
	40	47.9	25.8	29.6	18.6	10.5	10.1	24.5	3.3	8.9
8.2	10	40.7	28.3	28.1	15.3	10.2	9.3	23.2	7.5	11.3
	20	39.7	27.0	29.1	12.9	10.0	9.1	30.9	6.0	12.9
	30	49.1	32.3	27.1	18.1	12.3	9.6	24.7	6.6	8.9
	40	46.7	27.8	28.9	16.7	11.5	7.6	27.6	6.7	10.3

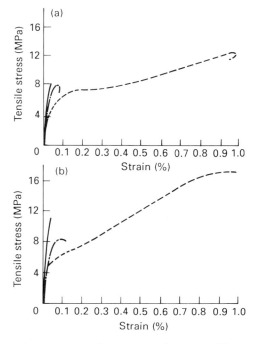

Fig. 4.9 Tensile stress–strain curves of grc composites stored in --- air, ⎯⎯ water and —·—·— natural weather for ten years. Fibre length 40 mm; fibre volumes (A) 4.4% and (B) 8.2%.

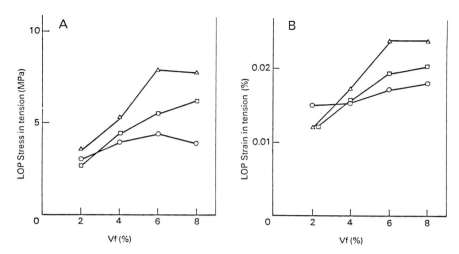

Fig. 4.10 Variations in the limit of proportionality (LOP) stress (A) and strain (B) in tension with fibre volume fractions at ten years. Composites stored in air (○), water (△) and natural weather (□).

unchanged with time when kept in relatively dry air. By contrast, in a completely wet environment a significant reduction in strength takes place. Under natural weathering conditions the MOR of the composites containing sand decreases with age and at 10 years no significant differences are noticeable in the strength of the composites with or without sand. The long-term effect of adding different proportions of sand on the bending and impact strength of grc in natural weather is illustrated in Fig. 4.11.

The incorporation of sand initially reduces the LOP and MOR values of grc due to the dilution of cement. It has been reported[7] that the LOP values of grc made from neat cement decrease with time when stored in a dry air environment, possibly due to the shrinkage of cement. The addition of sand to the cement paste reduces the drying shrinkage and that may allow the LOP values to stabilise when kept in a dry air environment. As can be seen from Table 4.3 grc composites containing up to 35% sand do not show any large change in their LOP values from those of composites made from neat OPC. The composites with 50% sand content indicate the long-term stability of LOP under dry conditions as there are no marked changes with age in these values. The grc composites containing sand when immersed in water or exposed to weathering conditions show a substantial improvement in their LOP values, reflecting the greater hydration of cement.

The direct tensile strength (UTS) of the composites shows no significant dependence on the proportion of sand used (Table 4.3) and the values are

Table 4.3 Properties of grc containing sand.

Matrix	Property	7 days Moist air	28 days Moist air	Age and storage environment 1 year			5 years			10 years		
				Air 40% RH	Water 20°C	Weather	Air 40% RH	Water 20°C	Weather	Air 40% RH	Water 20°C	Weather
80% OPC 20% sand	MOR (MPa)	33	36	34	24	30	31	19	22	31	18	19
	LOP (MPa)	14.5	15	13	18	18	9	17	14	11	16	12
	UTS (MPa)	—	12	12	8	11	11	4	8	11	5	7
	IS (kJ/m²)	20	—	21	8	14	19	4	8	18	3	6
	E (GPa)	—	29	23	38	34	23	32	27	32	30	25
	ε (microstrain)	—	8860	9610	470	4690	8200	150	570	9150	140	440
65% OPC 35% sand	MOR (MPa)	29	34	33	24	29	32	20	23	31	19	17
	LOP (MPa)	13	14	13	17	15	10	18	16	11	17	12
	UTS (MPa)	—	13	11	9	10	12	6	8	11	4	7
	IS (kJ/m²)	21	21	21	9	13	19	6	7	18	3	5
	E (GPa)	—	32	32	39	—	30	46	32	38	35	30
	ε (microstrain)	—	8060	6420	340	2300	8940	145	500	7920	146	280
50% OPC 50% sand	MOR (MPa)	25	28	33	26	31	29	19	22	29	18	18
	LOP (MPa)	11	12	14	16	17	12	17	15	12	17	15
	UTS (MPa)	—	12	10	9	9	9	9	7	9	4	7
	IS (kJ/m²)	23	21	17	8	12	14	4	10	14	3	3
	E (GPa)	—	40	34	36	37	24	32	46	36	35	43
	ε (microstrain)	—	9250	6540	315	3000	3510	350	240	3200	190	
50% OPC 50% sand 5% resin (on weight of total solids)							(6 years)					
	MOR (MPa)	15	28	27	16	23	26	14	20	—	—	—
	LOP (MPa)	—	12	11	13	13	9	12	12	—	—	—
	UTS (MPa)	—	9	10	7	7	10	4	7	—	—	—
	IS (kJ/m²)	—	24	24	4	12	18	3	10	—	—	—
	E (GPa)	—	—	23	23	22	27	29	25	—	—	—
	ε (microstrain)	—	9940	8770	400	2390	10200	217	1720	—	—	—

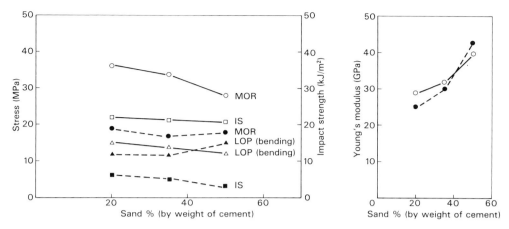

Fig. 4.11 Properties of grc composites containing sand stored wet in the laboratory for 28 days (open symbols) and on weathering site at Garston for ten years (filled symbols).

lower than those obtained with ordinary OPC/grc. Any change in the tensile properties with time depends on the storage environment. When kept in dry air, the UTS values remain essentially unchanged with age. However, under water storage a significant reduction in strength takes place. It is also clear from Table 4.3 that the UTS of grc composites with varying amounts of sand additions shows no significant difference when exposed to natural weathering conditions for ten years.

The Izod impact strength of grc composites containing sand also shows very similar results to that of ordinary OPC/grc. In relatively dry air, the impact strength of the composites is unaffected with age but in completely wet environments a serious deterioration takes place and the material becomes progressively brittle (Table 4.3).

Young's modulus of grc composites increases with increasing sand contents (Fig. 4.11). In the longer term the stiffness of the composites with added sand remains high when kept under water or natural weathering conditions. In a relatively dry air environment, the stiffness and the strain to failure in tension of composites containing up to 35% sand show no appreciable change with time up to five years, but are reduced considerably for grc containing 50% sand. There is an increase in Young's modulus values after ten years for all mix formulations.

The results on the grc board containing peridite resin given in Table 4.3 show a distinct improvement in pseudoductility in dry air and natural weather over a period of six years. Polymer modified grc is discussed more fully in Chapter 6.

4.1.1.4 CEM-FIL 2/GRC

The above results clearly indicate that the strength properties of grc made from OPC and Cem-FIL AR glass fibres with or without sand are reduced with age in wet environments. To overcome this deficiency more alkali-resistant fibres such as Cem-FIL 2 and AR fibre super[8] have been developed. The properties of grc made from Cem-FIL and Cem-FIL 2 are compared in Table 4.4. A significant improvement in the performance of grc in natural weather is achieved when Cem-FIL 2 fibres are used as reinforcements. Up to at least nine years of natural weathering at BRE, Cem-FIL 2/grc has remained very strong and tough.

Ferry[9] has made a detailed study of the effect of glass fibre length and content on the properties of Cem-FIL 2/grc containing sand. From accelerated ageing test results he concludes that for sprayed grc longer fibres should give higher strength in the composite in the long term. This observation is at variance with the real-time strength results of Singh and Majumdar[4] on Cem-FIL/grc discussed earlier. Ferry has attributed this difference to the superior durability of Cem-FIL 2 in cements compared to that of Cem-FIL. He also suggests that for effective reinforcement the glass content in sprayed grc should not fall below 3.5% by weight.

4.1.1.5 ACCELERATED AGEING TEST

Note that in Table 4.4 some results refer to curing of the composite for different periods of time in hot water at 50°C, and it is clear that under these conditions the reduction in the strength properties of grc takes place rather rapidly. This form of test, i.e. curing the composite in hot water at various temperatures, is now widely used in the grc industry as an accelerated ageing test, and the results given in Table 4.4 strongly suggest that a better long-term performance can be expected from Cem-FIL 2/grc compared with Cem-FIL/grc. A similar conclusion has been made about grc made from 'AR fibre super' manufactured by Asahi Glass in which the fibres are treated with a size that contains cement hydration inhibitors[8].

Combining the results of accelerated ageing tests on grc composites with the corresponding results of the Strand in Cement (SIC) tests described in Chapter 1, Litherland and his colleagues[10] have developed a methodology by which the long-term strength of grc containing AR glass fibres, such as Cem-FIL and Cem-FIL 2, can be predicted. The methodology is based on the assumptions that (a) the rate of loss in the glass fibre strength in cement is directly related to the rate of some chemical reaction, (b) the time taken for the SIC strength to fall to any given value σ_{sic} can be regarded as an inverse measure of the rate of strength loss, and (c) grc composite strength $\sigma_c = \sigma_f V_f$. An Arrhenius-type relationship, $\log_{10}$ (time to reach σ_{sic}) $\propto \frac{1}{T}$, where T is the absolute temperature of the accelerated test, has been found

Table 4.4 Properties of grc made from Cem-FIL and Cem-FIL 2.

Environment	Cem-FIL 2/OPC grc (70% OPC, 30% sand matrix)						Cem-FIL/OPC grc (70% OPC, 30% sand matrix)					
	MOR (MPa)	LOP (MPa)	IS (kJ/m²)	UTS (MPa)	E (GPa)	Failure strain (microstrain)	MOR (MPa)	LOP (MPa)	IS (kJ/m²)	UTS (MPa)	E (GPa)	Failure strain (microstrain)
28 day damp cure in laboratory	38	12	21	15	30	11000	36	18	19	14	38	8250
Time in hot water (50°C)												
10 days	36	9	16				28	17	8			
40 days	33	12	11	11			17	15	3 (after 53 days)			
90 days	28	14	8				13		2			
180 days	24	12	8	9	33	870	13	13	2 (after 223 days)			
1 year							13	13	1.3			
2 years	16	15	3	5	40	120						
Natural weathering at BRS												
180 days							35	17	18	13	38	5500
1 year							31	16	12	10	38	
2 years	36	11	20	13	39	8000	28	18	8	10	37	430
4 years	33	15	14	13	36	7310	22	15	5 (5 years)	8	41	240
9 years	30	14	13	11	44	4710	18	16	6 (10 years)	6	39	160

to be applicable to grc made from both Cem-FIL and Cem-FIL 2 AR fibres giving approximately the same value, 88–89 kcal/mole, for the activation energy for the overall strength reduction process[11].

At first the SIC strengths of the glass fibre strands of interest need to be determined for various temperatures and at different ages. The results obtained by Litherland et al.[10] for original Cem-FIL are shown in Fig. 4.12. The Arrhenius plots were then constructed by these authors from the data covering the time taken to reach SIC strengths from 1000 MPa to 300 MPa in experiments conducted at temperatures between 4 and 80°C (Fig. 4.13). As the straight lines in Fig. 4.13 are all parallel to each other, they can be 'normalised' with respect to any selected temperature T to give an overall picture of the relative acceleration of strength loss at different temperatures. Fig. 4.14 shows this 'normalised' plot for 50°C. Bend test data from accelerated as well as naturally weathered grc specimens from various locations in the world are plotted in the same figure. All grc composites shown in Fig. 4.14 were from spray-dewatered boards nominally containing 5 wt% Cem-FIL fibre, and for the weathered specimens the mean annual temperature of the location is plotted.

It is easy to see that the temperature coefficients given in Fig. 4.13 can be used for making quantitative predictions using time transpositions. Thus it has been suggested[11] that in the UK:

1 day at 80°C in water ≡ 1672 days (4.6 years) weather
1 day at 70°C in water ≡ 693 days (1.9 years) weather

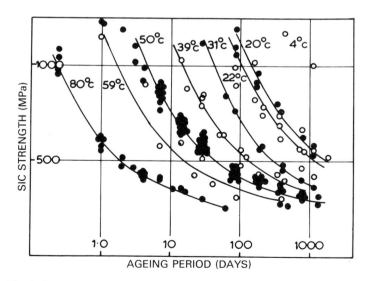

Fig. 4.12 SIC strength retention in water at various temperatures (Reference 10).

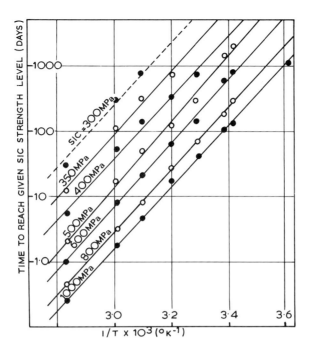

Fig. 4.13 Arrhenius graphs from SIC results (Reference 10).

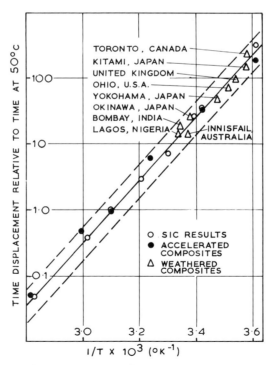

Fig. 4.14 Relative strength loss rates for SIC and composites in accelerated ageing and weather exposure (Reference 10).

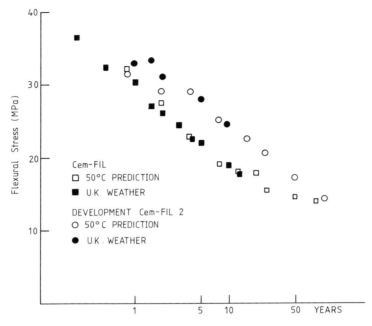

Fig. 4.15 Comparison between weathering results and predictions from accelerated ageing (Reference 12).

1 day at 60°C in water ≡ 272 days (9 months) weather
1 day at 50°C in water ≡ 101 days (3.5 months) weather

Using these equivalent factors the bending strength of grc made from AR fibres and kept in water at 50°C for 180 days is likely to be of the same order as that reached in typical UK weather (mean annual temperature 10.4°C) after 50 years. The predicted values of the long-term bending strength of grc made from Cem-FIL and Cem-FIL 2 fibres are compared with experimental values obtained to date in Fig. 4.15[12]. The agreement is very good.

Attempts are currently being made to develop predictive relationships for other strength properties of grc, e.g. impact strength[12].

4.2 Premixed grc

Although premixed grc is used extensively in manufacturing products by such methods as casting in open moulds, pumping into closed moulds, extrusion and pressing, data on the properties of such a composite material are very scarce in the literature. At BRE, Hills[13] studied the bending and

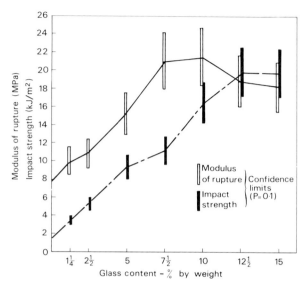

Fig. 4.16 Relationship between strength and glass content for premixed grc (OPC/Cem-FIL) at 28 days.

impact strengths of pre-mixed grc composites varying the glass content and his results are shown in Fig. 4.16. Premixed grc, like the spray-dewatered product, also exhibits an optimum limit of fibre addition above which the strength properties deteriorate, presumably because of increase in the porosity of the material.

For vibration-cast premix grc used commercially and having a sand:cement ratio of 0.5:1 and a water:cement ratio of 0.33, the initial mean property values are as follows[1]: compressive strength 40–60 MPa, MOR 10–14 MPa, LOP 5–8 MPa, UTS 4–7 MPa, strain to failure 0.1–0.2% and impact strength 8–14 kJ/m^2. Some changes in these values on long-term weathering are to be expected; it has been found that matrix controlled properties such as LOP increase but impact strength and strain to failure values are reduced[1].

4.3 Other properties of grc

The experimental values quoted in the following sections are those of Cem-FIL/grc. It is assumed that replacement of Cem-FIL by Cem-FIL 2 will only have a marginal effect on the properties described.

4.3.1 Density

The density of grc, like other properties, varies with the composition and method of fabrication. The normal ranges of dry bulk density are 2000 to 2200 kg/m^3 for spray-dewatered grc, 1900 to 2100 kg/m^3 for direct spray material and 1900 to 2000 kg/m^3 for premix grc.

4.3.2 Thermal expansion

Existing data suggest values $7-12 \times 10^{-6}/°C$ for the fully dry and fully wet grc. At intermediate moisture content, values up to $20 \times 10^{-6}/°C$ have been obtained in tests.

4.3.3 Thermal conductivity

This varies with the density between 0.5 W/m°C at 1700 kg/m^3 and 1.3 W/m°C at 2200 kg/m^3.

4.3.4 Shrinkage and dimensional movement

In common with other cement-based materials, grc undergoes drying shrinkage on exposure to low humidity. The initial drying shrinkage for the wet-cured condition to dry (the ultimate shrinkage) consists of an irreversible component and a moisture movement which is reversible in further wetting and drying cycles. For material with a neat cement matrix the ultimate drying shrinkage can be up to 0.4% at 50°C and 30–40% RH. Incorporation of sand in the mix reduces both the shrinkage rate and the ultimate shrinkage, which is about 0.15% for a mix with a cement/sand ratio of 1:2. The reversible component, the moisture movement, is usually about two thirds of the initial drying shrinkage.

On-site dimensional changes due to temperature fluctuations are superimposed on those resulting from changes in moisture content. Measurements of the total dimensional changes due to changes in moisture content and temperature of small grc wall and roof panels on the BRS exposure site at Garston indicate that the overall movements of grc components used externally in temperate climates are likely to be less than the ultimate values of drying shrinkage measured in the laboratory.

Unlike artificially dried material there is evidence that the maximum overall shrinkage from wettest winter weather to driest summer weather may increase as the seasonal cycles progress.

4.3.5 Permeability

The permeability of grc to both air and water varies with its compaction and density, and the storage conditions before test. The following observations have been made[14]:

(a) Air permeance increases with storage in dry conditions: with 9 mm thick grc, a value of 3 metric perms was recorded after one year at 40% RH, and this increased to 3.3 metric perms after two years. Values below 0.15 perms were obtained for material stored under wet conditions.
(b) The water vapour permeance of grc is dependent upon age, storage conditions and initial water/cement ratio. In general the permeance decreases with age. The following values were determined by the wet cup method, with 9 mm thick sheets.

Water/cement ratio	Water vapour permeance (metric perms)		
	At 28 days	After 2 years	
		In water	Weathering
0.25	3	0.6	0.8
0.35	5	1.7	2.1

4.3.6 Fire

Provided it does not contain more than a small amount of organic substances grc is non-combustible to Part 4 of BS 476 and does not propagate flame, and therefore satisfies all the requirements of Class O as defined in approved documents B 2/3/4.

The fire resistance of an element of structure is defined by its performance to the relevant part of BS 476, i.e. load-bearing capacity, integrity and insulation.

The performance in relation to these criteria of an element of structure incorporating grc varies with the formulation, method of manufacture and thickness of the grc, and also with the type and thickness of any core material in a sandwich construction. A single skin of the normal cement/sand mix, for example, will not satisfy the insulation requirement and cannot be relied on to maintain integrity. Fire-resistant grc mixes have been developed that incorporate lightweight aggregate and/or entrained air bubbles, and these maintain integrity in single skin form.

GRC sandwich panels with suitable lightweight cores, such as 'Styropor' concrete, perform well in the fire-resistance test, satisfying all three criteria with ratings of up to four hours.

4.3.7 Creep and stress rupture

Measurements of creep in tension of spray-dewatered Cem-FIL/grc have been made for indoor and water storage conditions, and for specimens kept out of doors. The immediate deformation on loading the grc is followed by a further slow creep deformation, with the creep rate decreasing with time. Below the LOP in tension the creep strain is small, and is quantitatively the same as that of the unreinforced matrix. Even so the creep strain exceeds the elastic strain after about three weeks and after two years is 2 to 3 times the (elastic) strain[1].

Out-of-doors dimensional changes due to humidity and temperature changes are superimposed on the creep strains. These dimensional changes can be large and at normal working stress levels are likely to be much more important than strains due to creep[15].

There have been no comparable stress rupture tests in tension, but in sustained loading tests in bending, no stress ruptures have been observed at up to twice the normally recommended levels of working stress, for tests under water, in dry air and out-of-doors (i.e. UK weather) over a period of a few years[16].

4.3.8 Fatigue behaviour

In repeated load fatigue tests in bending and direct tension on spray-dewatered Cem-FIL/grc, stored for up to one year, a common form of S/N curve (stress against cycles to failure) is obtained[17]. The bending tests gave median fatigue lives greater than 10^5 cycles at the LOP stress level and greater than 10^6 cycles at normal flexural working stress levels (6 MPa); the storage condition had little effect on the median lives. Direct tension tests indicated lives in excess of 10^4 cycles at the BOP and over 10^6 cycles at normal working stress levels, for material stored in wet and in dry conditions for two years.

After prolonged storage in water (six years) the resistance to flexural fatigue decreased: at normal working stress levels in bending (6 MPa) the median life was still about 10^6 cycles but at stress levels near the LOP it was only about 2×10^3 cycles[18].

4.3.9 Freeze−thaw behaviour

Artificial (laboratory) tests of the freeze−thaw behaviour of grc have shown relatively little change in properties, LOP increasing by up to 20%, while MOR, modulus and impact strength decreased by up to 20% in the more severe tests. However, these tests do not duplicate natural exposure, and interpretation is therefore difficult.

In natural freeze−thaw conditions (in Toronto, Canada) no sign of damage is visible in panels exposed for up to ten years[1].

4.3.10 Chemical resistance

The chemical resistance of grc made from Portland cement is similar to that of well-made concrete made from the same cement. There is little evidence to suggest that AR glass fibres such as Cem-FIL are affected significantly in sulphate solutions or in contact with dilute acids under conditions that are not detrimental to the cement matrix. The alkali resistance of these and similar fibres has been discussed in detail in Chapter 1.

Spray-dewatered grc is less permeable to CO_2 than concrete and under some conditions carbonation that may take place is likely to be limited to the surface zone. The chemical resistance of grc has been discussed in a general way by Warrior and Rothwell[19]. In particular situations, e.g. marine applications, the performance of grc may be improved by the use of special cements.

Chapter Five
Grc From Modified Portland Cement Matrices

Grc components are rarely made from neat cements; for controlling shrinkage cracking it is necessary to incorporate a quantity of sand in the matrix. The strength properties of grc containing various proportions of sand are described in the preceding chapter. The other main object of making suitable additions to cement is the reduction of the overall alkalinity of the matrix. If this can be achieved it is expected that the composites will retain, in the long term, a much larger proportion of their very good initial strength and toughness. Some materials, when added to cement, react with $Ca(OH)_2$ produced during the hydration of cement. As the precipitation of $Ca(OH)_2$ crystals and their subsequent growth inside the glass fibre bundles is an important factor in controlling the long-term properties of grc, the reduction in the amount of $Ca(OH)_2$ in grc by reactions with suitable pozzolanas would be advantageous.

By including low-density materials in the matrix, lightweight grc with good insulation properties can be produced.

The effects of various types of additions on grc properties were studied at BRE and the results are summarised in this chapter. Unless otherwise mentioned, the properties refer to those of spray-dewatered grc having approximately 5 wt% of 32−34 mm long Cem-FIL AR glass fibres.

5.1 Fillers

Several inorganic materials used in grc as cement replacements, e.g. pulverised fuel ash (fly ash) or ground granulated blast furnace slag, enter into a pozzolanic reaction with the cement matrix and are of great interest. The properties of grc containing these materials are discussed in some detail in a later section. Among the non-pozzolanic fillers, the effects of including 20% Fuller's earth and 40% China clay waste or quarry fines in grc formulations as replacements for cement have been investigated[1]. The properties of these composites, kept in various environments for up to eight years, have led to the conclusion that these fillers act mainly as cement diluents. Matrix controlled properties of grc such as LOP and Young's modulus are reduced by the incorporation of these fillers. The initial strength of the composite is also reduced but the long-term strength is hardly affected. In this respect non-pozzolanic fillers do not contribute to the durability of grc, although like sand they may reduce the drying shrinkage of the composite. By optimising the initial cure of the composite (e.g. under water at elevated temperatures) some fillers can probably give sufficiently high early strength to be of interest in grc manufacture.

5.2 Pozzolanas

Pozzolanas, i.e. inorganic silicate materials that react readily with $Ca(OH)_2$ produced by the hydration of Portland cements, are finding increasing application as a cement diluent on economic as well as energy saving grounds. Many of the naturally occurring materials, volcanic tuff, for instance, are pozzolanic. The name pozzolana derives from a natural material found in Pozzuoli near Naples in Italy which had been first used many centuries ago in making the early Roman mortar. Many by-products of the modern industry, pulverised fuel ash from power stations or silica fume from the ferrosilicon industry, for example, are pozzolanic in nature and react easily with the product of cement hydration.

In general, pozzolanas have been found to be useful additions in grc.

5.2.1 Natural Pozzolan

So far, the effect of only one material of this type has been studied in depth in the evaluation of the long-term properties of modified grc[1]. The material is from Italy and described as 'Pozzolana Di Salone'. It is of volcanic origin, of large surface area (8240 cm^2/g), and consists of glass intermixed with quartz, feldspar, zeolites and pyroxenes. The chemical analysis is given in Table 5.1. The properties of spray-dewatered grc made from a matrix of 60% OPC and 40% pozzolana with 5 wt% Cem-FIL glass fibres are shown in Table 5.2. The composites were cured in two different environments up to ten years. Comparison of these results with those from standard grc (Chapter 4) indicate that in wet conditions one could expect higher long-term bending and impact strengths in the product containing the pozzolana but the LOP of this product is likely to be much lower than that of standard grc. The beneficial effect of the pozzolana is thought to be due to the formation of C−S−H at the expense of crystalline $Ca(OH)_2$. This results in

Table 5.1 Chemical analysis of pozzolanas.

	Italian Pozzolana	Pulverised fuel ash (Castle Donnington)	Ground granulated blast furnace slag (Cemsave)	Microsilica (Silica Fume)
SiO_2	44.51	49.3	33.2	96.82
Al_2O_3	16.26	24.8	13.6	0.23
Fe_2O_3	9.70	9.6	0.43	0.02
MgO	4.03	2.13	6.57	0.37
TiO_2	0.80	0.82	0.41	—
CaO	10.29	3.35	41.0	0.09
Na_2O	1.70	1.17	0.31	0.11
K_2O	5.73	3.88	0.54	0.32
Mn_2O_3	0.20	0.07	0.65	0.02
SO_3	0.003	0.62	—	0.16
P_2O_5	—	0.21	0.03	0.05
Cl	—	0.07	—	0.03
Total S	—	—	1.52	—
LOI	6.60	3.03	—	1.15

Table 5.2 Properties of grc containing Italian pozzolana.

Matrix	Property	Age and storage conditions						
		28 days		1 year		5 years	10 years	
		Air 40% RH	Water	Air 40% RH	Water	Water	Air 40% RH	Water
60% OPC, 40% Italian pozzolana	MOR (MPa)	41	35	40	34	28	31	26
	LOP (MPa)	10	12	9	12	10	7	11
	UTS (MPa)	—	15	—	15	—	—	—
	IS (kJ/m^2)	26	21	22	12	9	20	5
	ρ (g/cm^3)	1.96	—	—	—	—	—	—

a relatively porous interface but the severity of the OH$^-$ ion attack on the glass is also restricted as the concentration of these ions is reduced.

5.2.2 Pulverised fuel or fly ash

It is well known that several important benefits accrue from the addition of pfa (or fly ash) to cement, among them lower heat evolution, reduced permeability and porosity and lower drying shrinkage. As these factors are directly related to the durability of cement products pfa is of interest to the grc industry. Furthermore, as pfas are known to react readily with the products of cement hydration, notably $Ca(OH)_2$, they would be expected to reduce the degree of strength reduction in ordinary grc with time in wet environments. Consequently, a great deal of attention has been paid to the properties of grc containing pfa.

The properties of grc made from Cem-FIL AR glass fibres and containing various proportions of pfa and kept in three different environments up to ten years[2,3] have been studied at BRE and are listed in Table 5.3. A limited number of results on one board after 17 years in water and natural weathering is also listed in Table 5.3. The chemical composition of one of the pfas used is given in Table 5.1. It may be noted that Pozament is a commercially blended product of Pozament Cement Ltd, Boston Spa, West Yorkshire, containing approximately 50% pfa. It should also be pointed out that not all the composite boards in Table 5.3 could be made at the same water/solid ratio. This produced a variation in the density of the boards, from 1.54 to 1.91 g/cm^3, resulting in different fibre volume fractions in different composite boards. The fibre proportions in the composites in Table 5.3 were 3.76 vol % for the neat OPC board but only 2.91 for the 50% OPC 50% pfa board. As fibre volume is perhaps the single most important parameter in reinforcement by fibres, such a variation makes the direct comparison of results inadmissible. Furthermore, as pfa from more than one source and three different batches of cement were used, the matrices, even if the composites had the same density, were not the same in each case.

Nevertheless, certain trends in grc properties resulting from pfa addition are discernible from the results in Table 5.3. The incorporation of pfa in grc reduces the bending strength of the composite at early ages, the degree of reduction increasing with the proportion of pfa used. The limit of proportionality (LOP) values of grcs containing pfa are also lower than those of their neat OPC counterpart. With the passage of time the MOR of pfa containing grcs is reduced, the extent of the reduction depending on the environment of use and the amount of pfa in the composite. Generally speaking, the higher the proportion of pfa in the mix the lower the rate of decrease in MOR, and there is some suggestion that for a 50% replacement of cement by pfa the MOR values remain reasonably constant in wet environments at least up to nine years. As with ordinary grc (Chapter 4), those containing pfa show the least reduction in MOR in a relatively dry environment.

Again, like ordinary grc, those containing pfa retain their high initial impact strength for up to ten years when the composites are kept in relatively dry air, but in wet conditions a serious deterioration is observed and the material becomes progressively brittle, the effect decreasing with an increase in the amount of pfa used.

The properties of the composite that depend essentially on the matrix,

Table 5.3 Properties of grc containing different proportions of pfa.

Matrix	Property	28 days			5 years			10 years			17 years	
		Air 40% RH	Water	Natural weather	Air 40% RH	Water	Natural weather	Air 40% RH	Water	Natural weather	Water	Natural weather
OPC	MOR (MPa)	35–50*	31	—	30–35	21–25	21–23	31–39	17–18	15–19		
	LOP (MPa)	14–17*	12	—	10–12	16–19	15–18	14–16	16–17	13–16		
	UTS (MPa)	14–17*	13	—	13–15	9–12	7–8	11–15	6–8	7–8		
	E (GPa)	20–25*	—	—	20–25	28–34	25–32	25–33	25–31	27–30		
	IS (kJ/m²)	17–31*	23	—	18–21	4–6	4–7	15–22	2–3	2–6		
85% OPC	MOR (MPa)	36	—	29	27	17	14	23	15	6		
15% pfa	LOP (MPa)	11	—	11	9	12	11	11	12	6		
(CD)	UTS (MPa)	14	—	14	8	6	4	7	6	2		
	E (GPa)	—	—	—	23	32	—	25	36	15		
	IS (kJ/m²)	26	23	22	18	4	4	14	3	2		
	ϵ_μ (microstrain)	—	—	—	8100	250	330	3700	230	170		
75% OPC	MOR (MPa)	33	29	29	26	23	17	27	17	10		
25% fpa	LOP (MPa)	10	10	10	9	12	11	10	12	8		
(CD)	UTS (MPa)	14	13	13	11	9	6	11	8	4		
	E (GPa)	—	—	—	22	27	24	30	32	20		
	IS (kJ/m²)	24	18	20	23	6	6	19	2	3		
	ϵ_μ (microstrain)	—	—	—	11320	530	430	11650	470	320		
60% OPC	MOR (MPa)	31	28	—	28	26	17	28	21	14	13	12
40% pfa	LOP (MPa)	8	7	—	9	12	10	8	12	10	10	10
(WT)	UTS (MPa)	—	—	—	—	—	7	—	9	5	5	5
	E (GPa)	—	—	—	—	—	21	—	23	17	23	19
	IS (kJ/m²)	19	16	—	17	7	5	20	5	4	2	3
	ϵ_μ (microstrain)	—	—	—	—	—	480	— (11 years)	620	450	—	—

continued

Table 5.3 Properties of grc containing different proportions of pfa continued.

Matrix	Property	28 days			5 years			10 years			17 years	
		Air 40% RH	Water	Natural weather	Air 40% RH	Water	Natural weather	Air 40% RH	Water	Natural weather	Water	Natural weather
50% OPC 50% pfa (CD)	MOR (MPa)	24	17	—	—	20	19	—	17	13		
	LOP (MPa)	7	7	—	—	12	11	—	10	10		
	UTS (MPa)	—	7	—	—	—	—	—	7	4		
	E (GPa)	—	15	—	—	—	—	—	20	17		
	IS (kJ/m^2)	23	18	—	—	14	9	—	8	10		
	ϵ_μ (microstrain)	—	7060	—	—	—	—	—	460	430		
Pozament	MOR (MPa)	26	23	25	24	27	21	21	24	19		
	LOP (MPa)	7	8	8	7	10	12	7	10	10		
	UTS (MPa)	11	11	11	10	10	7	10	9	6		
	E (GPa)	—	—	—	10	23	19	15	27	18		
	IS (kJ/m^2)	20	17	20	21	14	11	18	12	11		
	ϵ_μ (microstrain)	—	—	—	18700	5420	2350	12130 (9 years)	3540	910		

* Total range for air and water storage conditions
(CD) = Castle Donnington pfa
(WT) = West Thurrock pfa

such as Young's modulus, or the LOP and the bend-over point (BOP), increase with age in natural weathering, reflecting a greater degree of cement hydration. The stress–strain diagrams show that an increase in the amounts of pfa in the matrix increases the failure strain of the composite. This increase is manifested in composites having higher proportions of pfa, showing greater impact resistance.

The long-term weathering behaviour of grc containing pfa made from Cem-FIL fibres as determined by measurements on small specimens and described above is confirmed by results on large (1 m × 1 m) vertical and nearly horizontal panels[3]. Compared with panels made from neat OPC, the vertical panel containing 40% pfa had a much higher failure strain (0.21%) and impact strength after nine years of weathering but substantially lower LOP stress and Young's modulus.

Grc composites produced from the commercially blended OPC and pfa mixture, Pozament, have given encouraging long-term results in all three environments studied. Even when kept under water continuously over a period of nine years, the failure strain of the material has remained relatively high (0.3%), and the same feature, although to a lower extent, is seen in the weathered samples. The reason for this superior behaviour is not clear. Scanning electron micrographs of the fracture surfaces of five year old grc made from Pozament (see Chapter 8) showed substantial fibre pull-out and very little evidence of fibre corrosion. The fibre/matrix interface also appeared to be less dense compared to that seen in grc made from neat OPC[2]. These factors contribute to the retention of pseudoductility by the composites over the long term.

When Cem-FIL 2 fibres are used in place of Cem-FIL, marked improvements in the properties of grc are attained, as can be seen from the data in Table 5.4. The results are averages of those obtained from the two boards made from a 60% OPC 40% pfa matrix and Cem-FIL 2 fibres. The nine year weathering results were obtained on one board only. The glass contents and the water/solid (w/s) ratios of these two boards were very similar – 5 wt% and 0.3, respectively. The results obtained after keeping these composites under water at 50°C for up to 90 days show that a very large proportion of the strength and ductility of the material will be expected to be retained for a fairly long time. The accelerated test data are most complete for 50 days at 50°C and the retention of >0.8% ultimate tensile failure strain by the composite under this condition (see Table 5.4) is very encouraging. If the equivalence factor suggested by Proctor et al.[4] for OPC/grc applies in this case, one would expect the results obtained after 50 days at 50°C to correspond to 15 years of natural weathering in the UK. The actual nine year strength results from small weathered coupons are somewhat lower than the predicted values. The alkali resistance of Cem-FIL 2 fibres has

Table 5.4 Properties of grc made from 60% opc, 40% pfa and 5 wt% Cem-FIL 2 fibre.

Property	Moist cure		Hot water (50°C)				Natural weather			
	28 days	10 days	37 days	50 days	90 days	180 days	1 year	2 years	4 years	9 years
MOR (MPa)	32.5	32.5	32.1	31.8	32.1	35.3	33.5	31.4	30.8	25
LOP (MPa)	10.2	8.4	9.1	10.7	10.6	10.3	11.4	10.4	11.2	10
IS (kJ/m^2)	22.6	17.8	18.8	19.2	19.3	21.2	20.6	16.3	16.2	12
UTS (MPa)	—	—	—	12.4	—	—	13.8	11.6	—	10
E (GPa)	—	—	—	24.2	—	—	17.7	21.8	—	23
ϵ_{tu} (microstrain)	—	—	—	8620	—	—	9410	7140	—	5110

been shown to be better than Cem-FIL, and the encouraging results listed in Table 5.4 can be explained in this way.

Since the initial strength properties of the OPC + pfa mixtures are inferior to those of the neat cement and the properties of the grc composite such as the LOP and Young's modulus are controlled by the matrix, attempts have been made to improve these properties by subjecting the composite to an initial cure at high temperature including steam curing. Using Cem-FIL 2 fibres as the reinforcement it has been shown[5] that if steam curing is carried out before the final setting of the OPC + pfa matrix, the properties of the grc composite are adversely affected. Steam curing after the final setting of the cement produces better results but these are inferior to those obtained with similar but ordinary cured grc products. It appears that the glass fibres are considerably weakened when they are exposed to the high temperatures in a wet cement environment, and this factor overrides any advantage that may be gained through an initial improvement of the strength properties of the matrix.

A commercial grc product, made from OPC + pfa mixtures using Cem-FIL 2 fibres as the reinforcement[6], was introduced several years ago. The details of the formulation of this product are not known but it is interesting to compare its properties (Table 5.5) with those of similar grc composites containing pfa discussed in this chapter. The properties are given for both 'longitudinal' (L) and 'transverse' (T) samples, and these refer to the direction of movement of the wet flat sheet during manufacture, the longitudinal direction being parallel to the direction of the board movement.

It is evident from Table 5.5 that a considerable amount of anisotropy is present in the commercial grc board in respect of all properties, reflecting perhaps the anisotropy in fibre distribution. Compared to the sprayed grc made from neat cement (Table 5.3) the matrix-controlled properties such as the LOP and Young's modulus are lower in the commercial product due to the dilution of the cement by pfa. The two-year results for composites kept under water show small improvements in these properties and the tensile failure strain values also remain satisfactory. The two-year natural weathering results on the same composites also show satisfactory performance in terms of bending and impact strength but the tensile failure strain value for transverse specimens, at ~ 0.1%, is lower than expected from accelerated tests carried out at 50°C (Table 5.5). Under the accelerated test conditions chosen the composites have retained significant proportions of their initial bending, tensile and impact properties.

Table 5.5 Properties of a commercial general-purpose grc board.

Curing condition	MOR (MPa)		LOP (MPa)		UTS (MPa)		Failure strain (microstrain)		E (GPa)		IS kJ/m²	
	L	T	L	T	L	T	L	T	L	T	L	T
Constant weight at 20°C 65% RH	25.0	15.6	7.5	5.8	9.6	5.0	9710	6770	16.8	14.2	11.6	7.4
In water at 20°C												
28 days	22.9	15.9										
1 year	23.2	14.7	7.6	6.8	9.5	5.5	9060	7320	18.1	15.7	7.8	6.0
											7.2	4.8
2 years	24.3	14.4	8.6	7.8	8.9	6.0	8440	5140	22.0	20.2	7.0	5.7
In water at 50°C												
7 days	23.4	14.7	7.0	5.6							9.2	5.4
180 days	21.7	13.5	7.7	8.0	8.3	4.9	6520	1290	18.5	18.7	6.1	4.2
1 year	20.2	12.9	8.5	7.3	7.7	4.8	5300	900	20.4	19.4	4.6	3.5
Natural weathering												
1 year	20.5	13.0	9.8	8.9	9.3	5.0	9100	2720	19.9	19.3	6.5	4.5
2 years	26.3	16.3	14.1	14.6	9.1	6.4	6250	1070	23.6	22.7	8.8	5.2

L = longitudinal, T = transverse

5.2.3 Granulated blastfurnace slag

Ground granulated blastfurnace slags (ggbs) of different kinds are used commonly as cement replacements in concrete (BS 6699). The materials are weakly hydraulic on their own but reactive in the presence of alkalis and lime released during cement hydration. Granulated slags are essentially glassy with small amounts of crystalline melilite and merwinite. The chemical analysis of one granulated slag, Cemsave, produced by Frodingham Cement Co in the UK, is given in Table 5.1. In common with pfa, the addition of ggbs to cement reduces the heat evolution in concrete, minimising the risk of cracking during cooling. It also reduces the permeability of the composite and alters the pore size distribution in the paste.

When a Portland blast furnace slag cement containing 20% granulated slag is used as the matrix and Cem-FIL AR glass fibres as the reinforcement, the grc composite shows similar durability trends with time as observed in the case of ordinary grc. If the proportion of slag is greatly increased, say, to 70% replacement of cement, and Cem-FIL 2 fibres are used as reinforcements, the resulting composite is much more durable[5]. The properties of such a composite are given in Table 5.6. The tensile stress–strain diagrams of the composite after accelerated ageing under water at 50°C for 50 days and two years are shown in Fig. 5.1. These results refer to composites that were made by spray-dewatering and contained approximately 5 wt% of 32 mm long Cem-FIL 2 fibres.

It is clear from the results in Table 5.6 that if a large proportion of cement is replaced by granulated slag, the matrix is far less hostile to the glass fibres, even at 50°C. After nine years in natural weathering the composite has retained a very large proportion of its initial tensile failure strain and has remained highly impact resistant. The same is true of the composite kept at 50°C in water for 50 days. The two-year trace in Fig. 5.1 also shows a very high tensile failure strain capacity, although this was attended by a wide fluctuation in load. It is possible that the fluctuations in the load were due to the disruption of the bond between the fibre and the matrix. It is interesting to note that even after such severe accelerated ageing, the glass fibre strengths were not reduced sufficiently for the 'critical' fibre volume fraction for reinforcement to rise above the nominal 4 vol% present in the composites. Otherwise, a single fracture mode of failure would have been observed.

One can conclude from these results that, as far as natural weathering is concerned, if there is a large amount of granulated slag in the matrix the Cem-FIL 2 fibres may retain their high tensile strength and strain capacity for a very long time and hence such formulations of grc composites are particularly promising.

Table 5.6 Properties of grc containing Cem-FIL 2 and ground granulated blastfurnace slag.

Environment	30% OPC, 70% Cemsave matrix, Board D6						30% OPC, 70% Cemsave matrix, Board E18					
	MOR (MPa)	LOP (MPa)	IS (kJ/m²)	UTS (MPa)	E (GPa)	Failure strain (microstrain)	MOR (MPa)	LOP (MPa)	IS (kJ/m²)	UTS (MPa)	E (GPa)	Failure strain (microstrain)
28 day damp cure in laboratory	35	11	21				35	8	21	13	22	9823
Time in hot water (50°C)												
10 days	42	12	24									
50 days	37	11	20	13	33	9500						
90 days	25	10	11				31	10	19	12	29	7950
180 days							29	8	16			
1 year							29	11	15	10	36	9330
2 years							26	12	13	10	33	6900
Time in water (20°C)												
1 year							37	13	22	13	28	11600
2 years							36	11	22	14		9270
Natural weathering												
1 year	36	10	22	14	19	10800						
2 years	36	15	19	15	23	9000						
4.25 years	32	11	18	13	23	8600						
9 years	–	–	–	11	27	6800						

104 *Glass Fibre Reinforced Cement*

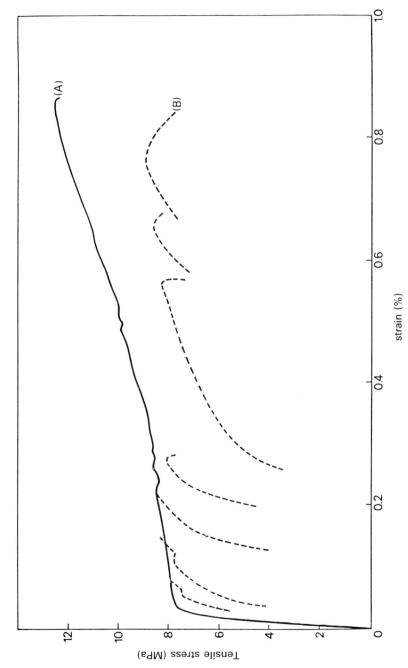

Fig. 5.1 Typical tensile stress–strain curves for grc made from Cem-FIL 2 fibres and 30% OPC, 70% slag: (A) age 50 days, (B) age 2 years; all stored under water at 50°C.

5.2.4 Silica fume

Condensed silica fume (microsilica) is collected in the form of ultrafine powders (typical specific surface area $\sim 25\,000$ m^2/kg) as a by-product in the manufacture of silicon and ferro-alloys. The material is strongly pozzolanic and the very small, spherical particles of microsilica provide significant filler effects when incorporated in concrete, reducing its permeability greatly. The resultant concrete is thought to be much more durable when compared with standard concrete, and hence microsilica has attracted a considerable amount of attention in recent years.

The effect of silica fume on the properties of grc is beginning to be studied systematically, although it is believed that in some commercial grc products a proportion of this material is included. A limited amount of work has been reported[5,7,8], and a summary of the results obtained by Majumdar *et al.*[5] on spray-dewatered boards containing silica fume and Cem-FIL 2 fibres is given in Table 5.7. The results correspond to two levels of substitution of cement by silica fume, at 20% and 40%. In order to produce a sprayable mix, superplasticisers have to be added to formulations containing silica fume. The amount of the superplasticiser increases with the amount of the fume and so does the water demand. The grc board where 40 wt% of the cement was replaced by silica fume had a final water/solids ratio of 0.43, much higher than the 0.30 common in standard grc manufacture.

For the board containing 20% replacement of the cement by silica fume, accelerated ageing produced a progressive reduction in the strength and toughness of the composite. In particular, the tensile failure strain was reduced from 1% after 28 days' damp cure to less than 0.1% after 180 days in water at 50°C. There was a related loss in the impact resistance of the material as measured by the Izod pendulum method. In general, Young's modulus of the composite and the limit of proportionality in bending increased with ageing. These are matrix-controlled properties and the increase is apparently due to the consolidation of the matrix brought about by progressive hydration of the cement and the reactions between silica fume and the products of cement hydration.

As manufactured, the grc board with 40 wt% replacement of cement by silica fume is weaker than the composite described above. This is due to the much higher w/s ratio (0.43) of the board. However, with the passage of time, particularly in a wet environment, there is an improvement in mechanical strength and stiffness of the composite confirmed by the accelerated ageing test results.

At the end of 180 days in water at 50°C, the MOR, LOP and UTS values measured for these composites are similar to those of the Cem-FIL 2/grc

Table 5.7 Properties of 70:30 opc/sand grc with part substitution of the cement by silica fume.

Environment	20 wt% substitution						40 wt% substitution					
	MOR (MPa)	LOP (MPa)	IS (kJ/m^2)	UTS (MPa)	E (GPa)	Failure strain (microstrain)	MOR (MPa)	LOP (MPa)	IS (kJ/m^2)	UTS (MPa)	E (GPa)	Failure strain (microstrain)
28 day damp cure in laboratory	28.4	8.4	20.1	10.7	22.4	10 081	19.2	5.4	19.9	7.8	15.7	10 250
Under water at 20°C												
28 days	28.7	8.7	19.5				23.1	6.9	19.2			
90 days	27.2	10.1	18.3				21.7	8.4	16.0			
180 days	24.7	10.6	15.9	9.9	29.9	9578	22.1	9.0	16.5	8.4	18.3	11 190
Under water at 50°C												
3 days	28.6	8.7	19.5				23.1	6.9	19.2			
7 days	28.7	8.6	17.2				24.4	7.0	20.8			
14 days	26.3	8.6	16.4				24.6	8.7	18.9			
28 days	23.3	11.1	12.4	9.1	35.3	7203	23.7	8.0	20.8	9.3	20.3	8483
56 days	21.1	7.4	13.2				21.6	11.9	16.0			
90 days	20.8	8.7	10.6	7.5	26.2	2908	21.9	11.7	18.9	8.5	28.4	7184
180 days	15.8	10.3	7.4	6.5	29.5	722	21.0	11.8	15.8	8.1	28.9	4151

made from OPC and sand (Table 4.4). Young's modulus of the composites modified by the inclusion of silica fume is still somewhat lower than that of the other kind but their toughness is much better. In fact the main advantage of adding silica fume to grc seems to be the more effective retention by the composite of its initial pseudoductility in the longer term. Silica fume will also have considerable influence on the permeability of the composite. If strength gains are desired from silica fume addition it appears that ways must be found for manufacturing the composite with a low w/s ratio. From the accelerated ageing test data given in Table 5.7 it would seem that grc made by using Cem-FIL 2 and containing significant proportions of silica fume is a material with useful long-term properties. It has also been shown[5] that dimensional changes such as drying shrinkage do not pose a serious problem for grc containing silica fume.

Hayashi et al.[7] have observed that if silica fume is added to the cement in the form of a slurry, the grc produced from the formulation has an initial strength similar to that of the product without silica fume. This advantage is expected to be retained at longer times as indicated by accelerated ageing tests. However, the degree of improvement in composite properties up to 20% silica fume addition was not outstanding. This is largely in agreement with BRE results. In laboratory experiments Bentur and Diamond[8] have attempted to incorporate silica fume directly into the spaces between glass fibre filaments in the strands. Tests of the behaviour of composites made from such fibres (and incorporating a further quantity of silica fume in the matrix) after accelerated ageing in water at 50°C for 28 days have indicated little loss in the strength and pseudoductility of the composite. It would be interesting to pursue this line of development in the commercial manufacture of grc.

Some interest has been shown recently in the use of metakaolin as a pozzolanic addition in grc. Metakaolin is produced from kaolin clay by calcination at high temperature. It reacts readily with $Ca(OH)_2$ like other pozzolanas, and this is believed to be the reason behind its effectiveness in improving the long-term strength properties of grc. Dejean et al.[9] have published results that demonstrate the effectiveness of metakaolin as an external addition in grc reinforced by A-glass fibres. Similar work on composites containing AR glass fibres is in progress in several laboratories.

5.3 Lightweight grc

Lightweight versions of grc are easily produced by incorporating a low-density filler in the composite mix. Perlite, bloated clay, pumice, etc. are examples of lightweight aggregates used in concrete. One most promising

Table 5.8 A comparison of one year old lightweight grc with commercial material.

Fibrous reinforcement	BRE material		Commercial material	
	Glass, 32 mm		Asbestos	Non-asbestos
Cem-FIL (vol %)	3.0	3.4		
OPC/cenospheres ratio (dry wt)	1.0	1.5		
Thickness (mm)	10	11	12	12
Density (kg/m^3)	940	1080	720	870
Bending strength (MPa)	12	14	14	11
Tensile strength (MPa)	4.4	6.7	4.4	4.0
Young's modulus (GPa)	9	9	2.6	3.6
Impact strength – Izod (kJ/m^2)	13	16	1	1
Thermal conductivity (W/m°C)	0.2	–	0.11	0.17

filler is the lightest fraction of pfa called 'cenospheres', which consist of hollow spherical shells providing closed porosity. The material is cheap and hence has received much attention. Lightweight grc is a suitable material for making insulation boards, and such products containing cenospheres are now made commercially.

Some of the properties of one-year old lightweight grc studied at BRE[10, 11] are compared in Table 5.8 with two other commercial materials. Grc sheets containing 20–60 wt% cenospheres when dried were produced by the spray-dewatering method using Cem-FIL AR glass fibres as the reinforcement and had dry densities in the range 700–1500 kg/m^3. The asbestos-containing material is no longer used in many countries because of the health hazards of asbestos.

Compared to the examples of asbestos insulation board, at one year the lightweight grc has similar bending and tensile strengths and a very much higher stiffness and impact resistance. Its thermal conductivity was greater than the asbestos board but tests have also shown that the force required to extract a No. 12 self-tapping screw from a 13 mm depth in the denser board is up to 1070 N, and that the less dense board, after five years' natural weathering, had a mean tensile strength perpendicular to the plane of the board of 2.3 MPa. These are useful properties for a lightweight material to have.

The measurement of the fire properties of lightweight grc shows its potential usefulness. Fig. 5.2 illustrates the unchanged appearance of a four-year old air-stored board 11.4 mm thick after a one hour fire test. Twenty minutes after the test had started, the mean temperature of the unexposed face had risen from 18°C to 375°C. Before the test the bending strength of the composite was 12.6 MPa, its LOP 6.4 MPa and its impact strength 12.4 kJ/m^2. After the one hour fire test, these values were 4.3 and

Fig. 5.2 Lightweight grc after one hour fire test (BS 476).

3.8 MPa and 1.9 kJ/m², respectively. As expected the fire-exposed face had weakened the most. It has also been shown that the thermal shrinkage undergone by lightweight grc is low.

Tests on the strength durability of lightweight grc composites containing cenospheres kept in different environments have also been carried out, and typical results for air storage are shown in Fig. 5.3. Similar results have been obtained with water-stored samples. It is seen that an increase in the proportion of cenospheres in grc weakens the products, but this has no effect on the change in its bending strength with time, which remains reasonably constant up to five years. Tests have also shown that the tensile strength follows a similar pattern. The impact strength of lightweight grc is reduced slightly with time in air storage, but when kept under water at 20°C a reduction in this property (e.g. from 11.7 kJ/m² to 7.0 kJm²) was observed

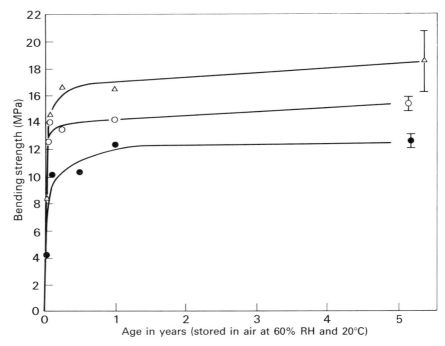

Fig. 5.3 Durability of air stored lightweight grc: OPC/cenosphere weight ratio. ● = 1.0, ○ = 1.5, △ = 1.9. Bars represent 90% confidence limits for the mean.

after five years. Strength retention in wet conditions is considerably improved when Cem-FIL 2 fibres are used as the reinforcement. For instance, the impact strength of lightweight grc containing nominal 5 wt% of Cem-FIL 2 fibres was found to be 12 kJ/m^2 when kept under water for five years. Other strength properties remained unchanged over the same period. It should also be remembered that the main uses of lightweight grc are likely to be inside buildings and therefore the long-term durability of the material in wet environments is a less important consideration here than with dense grc.

The lightweight grc can be easily nailed or sawn. The material can be used as the core for veneered panels since it can be bonded easily to other materials. It can then be considered a suitable substitute in most respects for various low-density boards containing asbestos.

Because of their greater porosity, lightweight grcs have poorer strength properties vis à vis their denser counterparts. This deficiency can be overcome, to a large extent, by 'impregnating' various polymeric substances into the composites. In a study[12], lightweight grc containing cenospheres was soaked in methyl methacrylate, then heat-cured. The polymer-impregnated lightweight grc with densities in the range 1000–1200 kg/m^3 has strength, stiffness and impact properties similar to or better than most grades of chipboard. The grc material will also withstand fire. When kept under water or exposed to natural weather for one year on the BRE site, lightweight grc with a dry density of 1150 kg/m^3 and containing 14 wt% polymer gave 10–11 MPa as the direct tensile strength and a failure strain of nearly 0.9% in some experiments. The corresponding bending strengths were 26–30 MPa, impact strengths 20–22 kJ/m^2 and Young's moduli 9–12 GPa. The results from accelerated ageing tests have also indicated that polymer-impregnated lightweight grc is likely to be more durable in wet conditions than the standard grc products made from Cem-FIL AR glass fibres. Polymer impregnation is, however, an expensive process and perhaps only in special circumstances could a material such as described here find application.

Chapter Six
Polymer Modified Grc

The properties of concretes and mortars made from Portland cements and incorporating dispersed polymers were investigated by many workers in the early 1960s and some of these results were discussed in detail at a RILEM Symposium[1]. The newer technique of polymer impregnation of concrete in which a liquid monomer is first allowed to permeate hardened concrete products and then polymerised *in situ* was first developed in the USA in the late 1960s[2]. A working party of the Concrete Society produced a report[3] in 1975 describing the main features of the various types of polymer concrete materials and assessing their potential for commercial applications.

As far as grc is concerned, Biryukovich and his colleagues were the first to use polymer dispersions[4], experimenting with cement composites reinforced by E-glass fibres. Commercial products based on high-alumina cement and E-glass fibres and incorporating a proportion of an organic polymer dispersion were made in the UK in the early 1960s and are believed to have performed reasonably well.

Among the various advantages claimed for polymer additions, a better workability of the mix, the mouldability of the product, less water absorption in the product and improvement in the strength and durability of the composite are perhaps the most important. Polymer-modified grc composites (pgrc) can be manufactured by spraying without the need for dewatering. The incorporation of polymers usually lowers the stiffness of the cement matrix, which has the effect of reducing stresses due to the drying shrinkage. A shorter early cure of the composite is thus possible.

With the development of Cem-FIL AR glass fibres, a programme of work was started at BRE in the early 1970s that aimed to examine the properties of polymer modified grc made from Cem-FIL fibres. Several candidate polymer dispersions were chosen and at first the properties of the cement matrix phase modified by these polymers were studied in detail[5,6] and only the preliminary results on grc containing one or two of the most promising polymers such as polyvinyl propionate or styrene-acrylics were published[7]. The long-term properties of many of these polymer modified grc composites have now been determined[8–10], and a summary of these findings is presented in this chapter together with significant contributions from other sources.

6.1 Polymer modified AR glass fibre reinforced cement

In the BRE study, either OPC or a mixture of 60 wt% OPC, 40 wt% pfa was used as the matrix material and the effect of a large number of polymers on grc properties was assessed. The composite boards were fabricated in the form of flat rectangular sheets, 1.5 × 1.0 m and approximately 10 mm thick by the spray method. The proportion of 32 mm long Cem-FIL AR glass fibres in all boards was nominally 5 wt% and that of polymer solids 10% of the weight of cement or cement plus pfa. The compositions of some of the polymer modified grc composites studied at BRE are given in Table 6.1.

All boards were given an initial cure of seven days in moist air, during which time they were cut into 150 mm × 50 mm specimen coupons and in most cases this was followed by 21 days' storage in air at 60–65% RH and 20°C to allow the polymer to dry out and form a film. The specimens were then distributed to different storage conditions for long-term studies.

The physical and mechanical properties of the more promising polymer modified grc composites studied at BRE are given in Figs 6.1–6.6 and Tables 6.1 and 6.2. It should be noted that in some cases results up to 15 years have been obtained. The abbreviations for the different polymers used in the tables are explained in the key for the figures.

6.1.1 Density and porosity

The bulk density and porosity of mature composite products containing polymers lie in the range 1600–1900 kg/m^3 and 18–36%, respectively. The addition of pfa results in a reduction of the bulk density of the material, and there is evidence to suggest that unmodified grc composites containing pfa were more porous. Composites containing AS and VPVC polymers had the lowest porosity. This factor might have had a significant effect on the long-term properties of these composites.

6.1.2 Flexural properties

Fig. 6.1 shows that the long-term bending strength of grc containing different types of polymer dispersions was higher than that of the control

Table 6.1 Composition of polymer modified grc.

Polymer	Slurry w/c*	w/c*	Freshly demoulded board								Oven dried at 105°C after 2 years in air at 60–65% RH, 20°C		
			Weight per cent				Volume per cent			Bulk density (kg m^{-3})	Bulk density (kg m^{-3})	Density (kg m^{-3})	Porosity (%)
			OPC	PFA	Polymer solids	Glass	PFA	Polymer solids	Glass				
A	0.28	0.23	71	–	7.1	5.1	–	12	3.7	1900	1770	2620	32
AS	0.25	0.25	70	–	7.0	5.2	–	12	3.8	1900	1850	2260	18
AS	0.42	0.42	42	28	7.0	5.4	28	13	4.2	2000	1620	2110	23
VPVC	0.50	0.24	71	–	7.1	5.3	–	12	4.2	2050	1900	2400	24
VPVC	0.93	0.42	42	28	7.0	5.3	23	10	3.6	1700	1580	2220	29
PVDC8	0.34	0.34	66	–	6.6	4.3	–	10	3.1	1840	1790	2480	30
PVDC11	0.24	0.24	70[a]	–	7.8	4.8	–	12	3.8	~2050	~1900	–	–
SBR	0.34	0.27	68[b]	–	8.8	4.6	–	16	3.4	1880	1890	–	–
SBR	0.53	0.53	40	27	6.7	4.9	22	11	3.1	1640	1630	–	–
Control + pfa	0.87	0.52	44	29	–	5.1	26	–	3.5	1800	1580	2480	36
Control	0.49	0.26	76	–	–	5.1	–	–	4.2	2150	1930	2710	29

* Water/cement ratio includes the weight of pfa as part of the cement.
[a] OPC batch 793; [b] OPC batch 737. Other boards contained OPC batch 766.

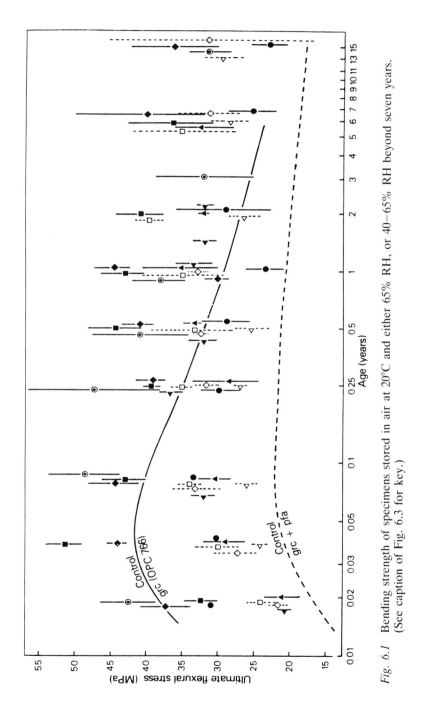

Fig. 6.1 Bending strength of specimens stored in air at 20°C and either 65% RH, or 40–65% RH beyond seven years. (See caption of Fig. 6.3 for key.)

when the composites were stored in relatively dry air. With some polymers, for instance SBR and acrylic, the strength of the composite at early ages was less than that of the control when neat cement was used as the matrix. With some other polymers, for instance AS, a small progressive reduction of strength with time has been observed. When the cement matrix was modified by the inclusion of pfa, polymer modification of the composite resulted in substantial strength gain at all ages. After 13 years the SBR modified grc containing pfa gave a bending strength value of ~ 30 MPa compared to a control value of ~ 18 MPa.

In continuous water storage (Fig. 6.2) the beneficial effects of polymer addition in grc noted above leading to an improvement of the long-term flexural strength of the product does not materialise, and after, say, about five or six years the strength of polymer modified grc is very similar to that of the control. This conclusion applies to composites containing pfa also.

Natural weathering (Fig. 6.3) results are more encouraging. Here polymer addition seems to have a definite positive influence as far as the retention of strength in grc over the longer term is concerned. After five or six years' exposure to natural weather at Garston, all polymer modified grc composites have given higher flexural strength values than the control: samples containing AS and VPVC giving outstanding results. The flexural strength value of ~ 22 MPa after 15 years of weathering obtained for grc modified by SBR, when compared with 15–19 MPa for control grc after 10 years,[11] illustrates the degree of improvement in the weathering resistance of grc that can be achieved by including a suitable polymer dispersion in grc formulations.

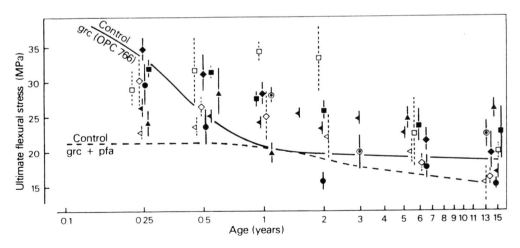

Fig. 6.2 Bending strength of specimens stored in water at 20°C. (See caption of Fig. 6.3 for Key.)

Polymer Modified Grc 117

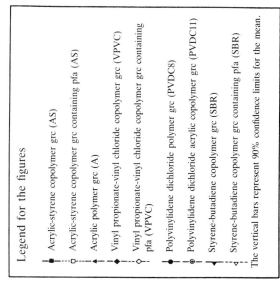
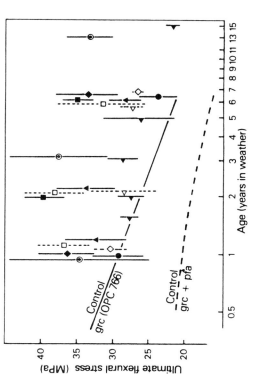

Fig. 6.3 Bending strength of specimens after natural weathering at BRE Garston.

118 Glass Fibre Reinforced Cement

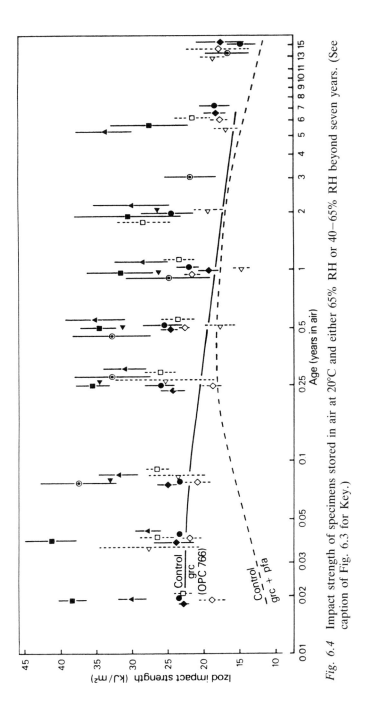

Fig. 6.4 Impact strength of specimens stored in air at 20°C and either 65% RH or 40–65% RH beyond seven years. (See caption of Fig. 6.3 for Key.)

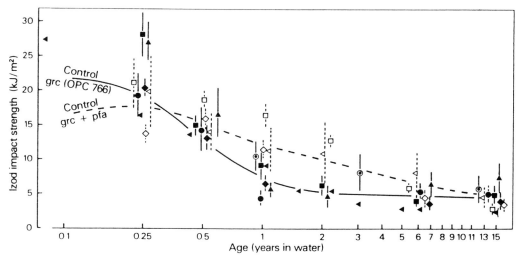

Fig. 6.5 Impact strength of specimens stored in water at 20°C. (See caption of Fig. 6.3 for Key.)

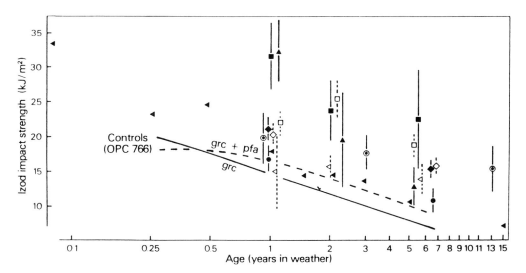

Fig. 6.6 Impact strength of specimens after natural weathering at BRE Garston. (See caption of Fig. 6.3 for Key.)

The flexural LOP values are also influenced by polymer addition, particularly at early ages when the composites are being cured. If the composites are kept in a relatively dry atmosphere, the incorporation of polymers such as AS or SBR can increase the LOP of standard grc by a substantial margin

Table 6.2 Properties of polymer modified grc.

Composite code and modifiers	Property	Unit	28–90 days		5–6 years			13–15 years		
			Air 65% RH	Water	Air 65% RH	Water	Weather	Air 40% RH	Water	Weather
35 A ▲	LOP UTS ε E	(MPa) (MPa) (microstrain) (GPa)	15 1 4400 —	16 — — —	17 11 960 15	24 — — —	22 12 670 21	— — — —	26 — — —	— — — —
25 AS ■	LOP UTS ε E	(MPa) (MPa) (microstrain) (GPa)	17 17 8800 27	17 — — —	22 16 >9000 20	23 10 400 25	22 13 5200 21	— — — —	23 — — —	— — — —
26 AS+pfa ☐	LOP UTS ε E	(MPa) (MPa) (microstrain) (GPa)	16 14 6760 16	13 — — —	15 15 5250 14	18 8 780 20	20 14 2430 16	— — — —	18 — — —	— — — —
27 VPVC ◆	LOP UTS ε E	(MPa) (MPa) (microstrain) (GPa)	13 18 11500 23	16 — — —	19 12 1400 26	20 3 100 28	19 11 800 28	17 — — —	19 — — —	— — — —
28 VPVC +pfa ◇	LOP UTS ε E	(MPa) (MPa) (microstrain) (GPa)	10 — — —	10 — — —	17 11 4950 20	11 6 390 22	15 7 890 22	16 — — —	11 — — —	— — — —
36 PVDC8 ●	LOP UTS ε E	(MPa) (MPa) (microstrain) (GPa)	13 — — —	13 — — —	17 8 1500 21	16 3 180 27	15 8 590 23	19 — — —	15 — — —	— — — —

Table 6.2 Properties of polymer modified grc continued.

Composite code and modifiers	Property	Unit	28–90 days			5–6 years			13–15 years		
			Air 65% RH	Water		Air 65% RH	Water	Weather	Air 40% RH	Water	Weather
96 PVDO 11 —◉—	LOP	(MPa)	23	–		–	–	–	24*	21	21
	UTS	(MPa)	15	–		–	–	–	–	–	13
	ε	(microstrain)	5000	–		–	–	–	–	–	700
	E	(GPa)	21	–		–	–	–	–	–	22
Pk,Pl SBR ▼	LOP	(MPa)	20	20		–	–	–	–	17	17
	UTS	(MPa)	15	13		–	–	–	–	–	–
73 SBR+pfa --▽--	LOP	(MPa)	10	11		12	12	15	12	16	–
29 Control GRC+pfa ----	LOP	(MPa)	7	8		10	10	10	14	12	–
	UTS	(MPa)	9	–		6	8	6	–	–	–
	ε	(microstrain)	8160	–		450	450	460	–	–	–
	E	(GPa)	12	–		22	26	22	–	–	–
30 Control GRC ——	LOP	(MPa)	13	16		14	17	15	14–16γ	17–19	13–16γ
	UTS	(MPa)	15	–		9	8	8	–	–	5
	ε	(microstrain)	8340	–		430	160	500	–	–	180
	E	(GPa)	32	–		36	–	37	–	–	32

γ –10 year old results from several grc boards[15]
* –65% RH 20°C

and the effect is equally pronounced for grc containing pfa. The difference between the modified and unmodified grc diminishes with time and in wet environments as the cement matrix hydrates and its strength is increased. Nevertheless, after five or six years, the contribution made by polymers such as A, AS and VPVC towards increasing the LOP values (Table 6.2) of grc kept in different environments remains positive and the effectiveness of polymer addition in this respect perhaps lasts for at least 15 years.

6.1.3 Tensile properties

Because of the difficulties inherent in determining the tensile properties of brittle materials such as grc after long-term weathering[12] there is uncertainty about some of the tensile test results listed in Table 6.2.

As manufactured, polymer modified grc including the products where a part of the cement has been replaced by pfa has a much lower value of Young's modulus than the control. This difference, arising from the much lower stiffness of the polymer compared to that of the cement, has lasted for at least five or six years. As would be expected, the difference is most marked for samples kept in air, where hydration of cement can proceed only slowly.

In general, as in the case with flexural strength, the long-term UTS values of polymer modified grc show little reduction when the composites were stored in air, much more in water storage, and weathered composites show an intermediate performance. The UTS of composites containing A, AS and VPVC after five or six years of natural weathering at Garston is much higher than that of grc made from neat OPC, and the composite containing AS as well as pfa has given very encouraging results. The two composites containing AS polymer, labelled 25 and 26 in Table 6.2, have also given high values for the failure strain after long-term weathering. Inasmuch as the tensile failure strain of grc has been identified as one of the key parameters governing the long-term viability of the material in applications such as cladding panels, the advantage of adding polymers in grc is obvious.

6.1.4 Impact strength

The impact strength results shown in Figs 6.4–6.6 parallel the flexural strength results regarding the effect of time and environment. In air storage, all polymer modified products are seen to be more impact resistant than the control, and they remain so for a long time. In water storage the polymer modified materials lose this advantage after a fairly short time — less than one year. But in natural weathering, some polymer dispersions such as AS and VPVC are still seen to be making a contribution to the impact resistance of the composite after five or six years. The 15 year results obtained with grc samples containing SBR (Fig. 6.6), however, indicate that in the very long term the impact resistance of polymer modified grc upon weathering may not be much better than that of standard grc.

The results given above clearly indicate that if suitable polymer dispersions are included in grc formulations, some of the properties of the composite are improved both in the short and long term depending on the environment. In the short term, the LOP of grc in bending is greatly improved and advantage can be taken of this fact, as pointed out by Ball[13] in shortening the initial cure of grc products. The long-term weather resistance of grc seems to be improved by the inclusion of polymers such as acrylic-styrene, vinyl proportionate–vinyl chloride copolymer or styrene butadiene rubber latex.

The contribution of the polymer in improving the weather resistance of grc cannot be fully assessed at the present time. Compared to neat cement paste, polymer cement composites are known to show low water absorption and greater strain capacity. The presence of a polymer may also improve the inter-particle bonding in the cement system and reduce stress concentration. Many polymer dispersions will form films around the cement grains and reduce their degree of hydration over a given period of time. The total alkali produced by cement hydration may thus be limited and consequently its corrosive effect on glass fibres may be substantially reduced. In a continuously wet alkaline environment, the polymer films are probably not stable and hence the long-term properties of the polymer modified materials do not show any improvement over their unmodified counterparts. In the course of weathering, dry cycles alternate with wet ones; this factor, together with temperature rises in summer, provide opportunities for the reformation of polymer films around cement grains and thus the advantages of polymer addition is not entirely lost.

If the above hypothesis is correct then by choosing an alkali-resistant polymer dispersion with a suitable film forming temperature and experimenting with the initial curing of the polymer modified grc composite it may be possible to improve the long-term strength retention of grc even further.

6.1.5 Cem-FIL 2/grc

Very recently Knowles and Proctor[14] and Bijen[15] have published detailed results on the properties of grc made by using Cem-FIL 2 fibres as the reinforcement and modified by the addition of different amounts of an acrylic polymer dispersion ('Forton' curing compound). Both studies emphasise the importance of curing in the development of the mechanical properties of polymer modified grc composites. Curing in warm damp conditions improves the LOP and MOR of the composite. Both studies agree that the incorporation of a suitable polymer in grc can result in significant reductions in drying shrinkage and moisture movement on subsequent exposure. These studies have also shown that for polymer containing grc there may be significant differences in the mechanical properties of samples tested in the 'wet' and 'dry' conditions. Knowles and Proctor draw particular attention to their observation that the LOP of the wet tested material is significantly lower than that of the dry material and suggest that this factor may be important in designing with grc in external applications. In Bijen's view[15] the difference in properties between the dry and wet tested polymer modified grc is mainly due to the swelling of the polymer in the wet material caused by water absorption. He found that the reductions in the LOP and MOR occurring when the composites were tested wet were fully recoverable on drying.

Long-term weathering results are not available at the present time for polymer modified grc made from Cem-FIL 2, but there is no reason to suspect that they would be any poorer than the corresponding Cem-FIL/grc composites (Figs 6.3, 6.6). Knowles and Proctor[14] have given some results of hot water ageing tests carried out on polymer modified Cem-FIL 2/grc composites which show relatively small differences between the polymer modified and ordinary grc. Both LOP and MOR of the polymer modified composites were a little higher than those of ordinary grc after prolonged ageing. Strain-to-failure values were similar after 56 and 84 days in water at 60°C. The authors suggest that for polymer modified grc these results are probably pessimistic estimates of weathering behaviour. There is also evidence to suggest that after long ageing periods the difference in properties between 'wet' and 'dry' tested materials becomes smaller and that the LOPs of wet tested materials improve significantly.

Knowles and Proctor[14], found no significant deterioration in the properties of polymer modified Cem-FIL 2/grc after the composites were exposed to ultra-violet radiation (ASTM G 53–84), freeze–thaw cycles (ASTM C 666), ozone at 40°C or after burial in soil at 25–27°C (BS 4618) for 12 weeks. It can, therefore, be concluded that, in general, the inclusion of a suitable polymer in grc formulations is beneficial. In this context it is worth mentioning that at present there is a considerable amount of interest in the grc industry in the so-called 5/5 pgrc mix. This formulation comprises 5% of 'Forton' polymer solids by volume and 5% by wt of AR glass fibres to be incorporated in grc. Daniel et al.[16] have shown that the seven days' wet curing normally recommended for grc is not required when the 5/5 mix is used for the development of comparable LOP values.

6.2 Polymer modified E-glass fibre reinforced cement

Following on from the work of Biryukovich et al.[4], Allen and Channer investigated the properties of polymer modified grc reinforced by E-glass fibres and published their findings in 1975[17]. They used an acrylic polymer emulsion, and Portland cement composites were made using a polymer solid/cement ratio of 0.2 in the slurry and chopped strand mats of E-glass fibre. Compared with plain glass cement composites, tests carried out 45 days after casting indicated that the polymer-modified grc material had good values of ultimate stress and strain in tension, about 14 MPa and 0.6%, respectively. The addition of polymer reduced the modulus of elasticity and creep was significant. The results also indicated that the polymer protects E-glass reinforcement from chemical attack by the Portland cement but no long-term weathering results were presented in support of this theory. In 1979 Bijen claimed[18] that, by combining a specially developed acrylic polymer and E-glass fibres the surface of which had been treated in a special way, grc composites with viable long-term mechanical properties could be produced. Based on this work a commercial product was launched in 1980 by Forton, a subsidiary of the Dutch Company DSM-Resins BV.

The strength properties of the E-glass reinforced polymer modified cement (Forton pgrc) including the durability aspects have been discussed in several publications and are summarised in Figs 6.7 to 6.12 taken from a paper by Jacobs[19]. The properties were measured on specimens taken from flat sheets (1.1 m × 1.1 m × 10 mm) made from the following formulation: cement: OPC; sand: fine silica sand; sand/cement ratio 0.2; glass fibre: Forton E-glass, 5% by volume; polymer: Forton polymer compound VF 765, 15% by volume; water/cement ratio: 0.3. The pgrc composite sheets were cured initially at 50°C for 16 h and then at 20°C and 65% RH for 28 days.

The results shown in Figs 6.7 and 6.8 suggest that E-glass pgrc composites retained large proportions of its initial bending and tensile strength after four years of natural weathering and the failure strain of the material in tension, ~ 1800 microstrain after four years of weathering, is higher than Cem-FIL/grc after similar exposure but much smaller than Cem-FIL 2/grc (see Table 4.4). The impact strength of the material (Fig. 6.9) has nearly halved during this period. The accelerated test results shown in Figs 6.10–6.12 show a significant retention of initial strength values and the tensile failure strain after six months in water at 50°C.

Bijen[20] has reported that, compared to the unmodified material, pgrc has a much lower moisture content at various relative humidities and its water absorption coefficient is also lower. Pgrc also has a lower shrinkage than the unmodified material. There is no indisputable evidence to suggest that fibre strength loss in the cement environment is reduced due

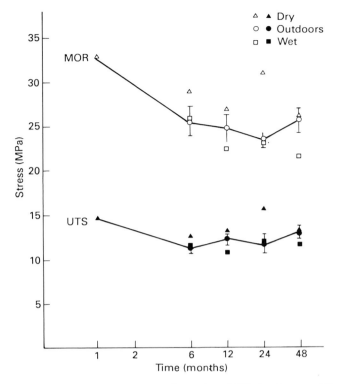

Fig. 6.7 Durability of pgrc, MOR AND UTS (Reference 19).

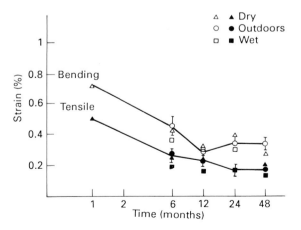

Fig. 6.8 Durability of pgrc, strain to failure (Reference 19).

Polymer Modified Grc   127

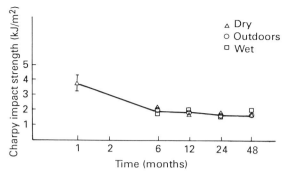

Fig. 6.9 Durability of pgrc, Charpy impact strength (Reference 19).

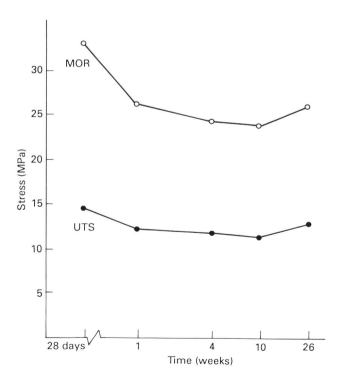

Fig. 6.10 Durability of pgrc, UTS and MOR; 50°C, under water (Reference 19).

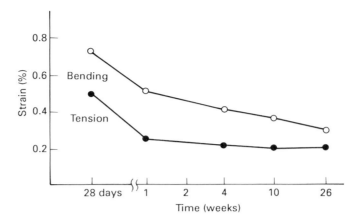

Fig. 6.11 Durability of pgrc, strain to failure, in tension and bending; 50°C, under water (Reference 19).

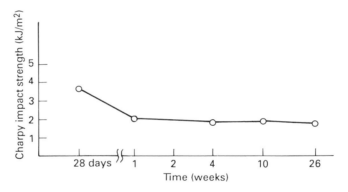

Fig. 6.12 Durability of pgrc, impact strength; 50°C, under water (Reference 19).

to the presence of the polymer. Work by Proctor et al.[21] indicates that polymer modification of E-glass/grc does not improve the long-term strength retention of E-glass fibres. Fibre surfaces after accelerated ageing tests of such composites showed evidence of gross chemical attack. Daniel and Schultz[22] have observed that unprotected E-glass fibres in polymer modified grc (Forton pgrc) show clear signs of alkali attack.

6.3 Polymer modified E-glass reinforced high-alumina cement

Allen[23] has studied the properties of HAC composites reinforced by four layers of E-glass chopped strand mat laid-up by hand. The composites were kept in different environments including water at 35°C and natural weather at Southampton.

Sixteen months' exposure out of doors caused a large drop in the tensile strength of the composite, from 15 to 9.3 MPa, and a corresponding decrease in strain to failure from 1.24% to 0.1%. Immersion in warm water caused even more drastic reductions so that the composites were no stronger or more pseudoductile than the original matrix. Immersion in cold water was almost as severe in its effects. Specimens kept indoors at normal temperatures and humidities showed little or no reduction in strength and strain capacity with time.

Similar results were reported by Grimer and Ali[24] for spray-dewatered composites made from HAC and chopped E-glass fibre strands.

6.4 Polymer impregnated grc

Some results on the effectiveness of polymer impregnation of glass reinforced cement composites were discussed in a preceding section dealing with lightweight grc. Briggs and Ayres[25] tried the technique on dense grc that was uniaxially reinforced with Cem-FIL AR glass fibres. The AR fibres were placed in the cement matrix by filament winding and the composite plates were impregnated with styrene-acrylionitrile monomer, and the latter was polymerised *in situ* by exposure to a dose of γ-irradiation at a rate of 0.1 Mrad per hour. The fibre content in the product was approximately 8 vol% and the polymer content 11 vol%.

Polymer impregnation produced a large improvement in flexural strength and failure strain over the control grc and induced failure to occur mainly in a tensile mode. The mean failure stress, 122 MPa, was more than two times that of the control at 52 MPa.

The LOP of the polymer impregnated material was 15.4 MPa compared to 6.7 MPa for the control. Short-beam inter-laminar shear strength tests gave shear strength values of 2.4 MPa and 6.4 MPa, respectively, for the unimpregnated and the polymer impregnated material. Water absorption measurements showed that the water accessible porosity was reduced to 6.6% after polymer impregnation from a value of 23% for the untreated material.

Although these results indicated very impressive improvements in grc properties that can be achieved by polymer impregnation the additional cost has so far prevented commercial exploitation of the idea. It would be necessary, however, to study the fire resistance properties of these composites thoroughly before recommendations for their use can be made.

Chapter Seven
Non-Portland Cement Grc

It has been pointed out in Chapter 1 that certain hydraulic cements used in the construction industry, e.g. high alumina cement (HAC) and supersulphated cement (SSC) produce a slurry that is less alkaline than that produced from Portland cements. Although the nature of the pore solutions in hardened products from these cements is not known, it is expected to be less alkaline than the pore solution in hardened OPC. The sensitivity of glass fibres to an alkaline environment manifests itself in a substantial reduction in the tensile strength. A considerable amount of research has therefore been carried out in assessing the potential of cements that are less alkaline than Portland cements as suitable matrices for reinforcement by glass fibres. It is interesting to note that much of the early work on grc by Biryukovich and his colleagues[1] was carried out on composites made from high alumina cement and E-glass fibres, and such a material was used in several countries including the UK in the 1960s before the advent of alkali-resistant glass fibres.

The properties of grc composites made from several cements other than the Portland variety using Cem-FIL glass fibres as the reinforcement have been studied at BRE in great detail over many years[2-4]. A summary of this work forms the main part of this chapter. It should be borne in mind that the composition of some of the cements described in this chapter, e.g. supersulphated cement, varies greatly from country to country and this, no doubt, will affect the results one may obtain for the corresponding grc composites.

The HAC used at BRE was that known by the brand name Ciment Fondu. Some supersulphated cements were formulated at BRE and others were obtained commercially. The Portland blastfurnace cement used contained approximately 20% slag.

At BRE, fibre composite boards of water/cement (w/c) ~ 0.3, approximately 10 mm thick and containing nominally 5 wt% of 32–34 mm Cem-FIL AR glass fibres, were produced by the spray-dewatering technique. After initial curing for seven days specimens (150 × 50 mm) from these boards were randomly distributed to various environments for long-term durability tests. The main storage conditions were: (1) air at 20°C, 40% or

65% RH, (2) water at 20°C, and (3) natural weathering on the exposure site at BRE. The mechanical properties of the composites were determined after specified times following standard BRE test procedures[5].

7.1 High-alumina cement (HAC) composites

As compared with ordinary Portland cement (OPC) HAC hardens more rapidly and its sulphate resistance is superior. However, under certain conditions of use it suffers a reduction in strength following the conversion of the hydrate phases $CaO.Al_2O_3.10H_2O$ and/or $2CaO.Al_2O_3.8H_2O$ to another, $3CaO.Al_2O_3.6H_2O$. The reduction in the strength of HAC concrete due to 'conversion' was accepted[6,7] as the principal cause of failure of beams in a number of buildings in the UK a few years ago, and the use of HAC in structural concrete is not now recommended in this country. In fibre reinforced cement composites used mainly in non-load-bearing applications the strength of the cement matrix, although very important, does not play the predominant role it has in structural concrete. Nevertheless the use of grc made from HAC has been very limited so far and with good reason.

From the modulus of rupture (MOR) values shown in Fig. 7.1 it is clearly seen that in relatively dry air the bending strength of grc made from HAC

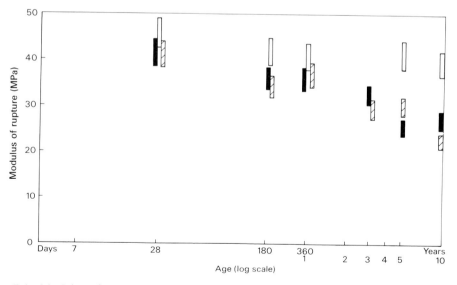

Fig. 7.1 Modulus of rupture of grc made from HAC and AR glass fibres: in air at 20°C, 40% RH □ ; in water at 20°C ■ ; under natural weathering ▨ .

and Cem-FIL remains unchanged with time at least up to ten years. When grc is kept under water or placed outdoors in the natural environment there is a reduction in strength, but the ten year values compare favourably with those of their OPC counterparts, for which MOR values of 17–18 MPa and 15–19 MPa under water and in natural weather, respectively, were obtained after ten years[8]. The impact strength of HAC/grc (Fig. 7.2) is also substantially better after ten years for the water-storage condition, although in natural weathering the long-term values are only marginally better than the 2–6 kJ/m^2 range measured for OPC/grc. The direct tensile strength results for HAC/grc after ten years are available only for the water storage condition, and this value, $\sim$ 8 MPa, is comparable to the 6–8 MPa quoted for the OPC composites. After ten years in water, Young's modulus of HAC composites was 30 GPa and their bending limit of proportionality (LOP) values were 13 MPa in air and water and 11 MPa in natural weathering.

After 20 years in water, composites made from HAC and Cem-FIL AR glass fibres have given the following results: MOR 15 MPa, LOP 10 MPa, UTS 7 MPa, E 35 GPa and IS 15 kJ/m^2. The corresponding values after 17 years of natural weathering at Garston were 17 MPa, 12 MPa, 7 MPa, 30 GPa and 8 kJ/m^2, respectively. The relatively high impact resistance of HAC composites at long ages is due to extensive fibre pull-out (see Chapter 8).

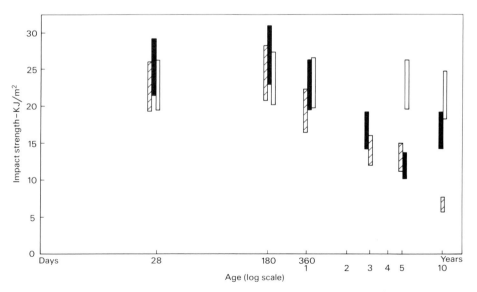

Fig. 7.2 Impact strength of grc made from HAC and AR glass fibres: stored in air at 40% RH ▯ ; water at 20°C ▮ ; under natural weathering ▨ .

At ambient temperatures and under wet conditions Cem-FIL/HAC composites thus appear to retain a much higher proportion of their initial strength in the long term than the corresponding OPC composites (see Table 4.1) reflecting the less alkaline nature of HAC[9]. It has also been shown conclusively[2] that under such conditions the Cem-FIL AR glass fibres are far superior as reinforcements to the HAC matrix than the borosilicate E-glass fibre.

It is well known that HAC loses strength rapidly at high temperature. Some idea of the effect of a rise in temperature on the properties of HAC composites can be obtained from the data in Table 7.1. In water the strength durability of the material is seen to be adversely affected by a rise in the temperature of the sample. Even when kept in the laboratory air at 35°C (a temperature likely to be reached on the surface of the specimen in the south-east of England in summer), there is a significant loss in strength with age. There are two main reasons for this increase in strength reduction:

(1) The AR glass fibres are corroded by the alkaline cement more at higher temperature, and correspondingly their strength is further reduced.
(2) The 'conversion' reaction that takes place in HAC leading to the reduction in the strength of the matrix proceeds at a much faster rate when the temperature is increased[10,11].

In grc, the matrix is not considered to be a very important direct contributor to the ultimate flexural or tensile strength of the composite because of its very small failure strain. But its integrity is important, and furthermore a progressive increase in the porosity of the matrix due to 'conversion' can affect composite properties by interfering with the stability of the interface between the fibre and cement. However, the results in Table 7.1 suggest that incursions into temperature regimes higher than ambient would not affect the ultimate strength of the composite provided that the increase in temperature is not excessive and is not sustained over a very long period of time.

Proctor and Litherland[12] have also studied the properties of grc made from HAC and Cem-FIL AR glass fibres over a period of several years. For composites containing 5 wt% fibre, i.e. similar to those studied at BRE, their results on composites weathered outside for ten years at St Helens and Fort Williams are given in Table 7.2.

The ten-year results of both types of composites show satisfactory retention of bending strength, and the impact strength data indicate that reasonable toughness has been retained after ten years of weathering. The increase in the LOP with time is also beneficial.

Table 7.1 Strength of grc made from HAC containing 4.7 wt% AR glass fibres at various temperatures.

Property	Temp (°C)	Curing condition	Strength (days)					
			7	15	28	90	180	360
MOR (MPa)	18	Water	35	33	35	32	30	26
	25	Water	—	—	34	30	29	25
	35	Water	—	32	34	29	25	25
	50	Water	—	29	28	24	20	—
	60	Water	—	26	27	21	—	—
	35	Air (40% RH)	—	40	39	36	33	28
Impact strength (kJ/m^2)	18	Water	48	41	38	37	35	21
	25	Water	—	—	42	36	20	17
	35	Water	—	38	42	31	21	19
	50	Water	—	26	19	14	14	—
	60	Water	—	19	18	16	—	—
	35	Air (40% RH)	—	39	46	36	32	33

Table 7.2 Properties of grc made from HAC and Cem-FIL (Reference 12).

Test time	Composite	MOR (MPa)	LOP (MPa)	Impact (kJ/m^2)
Start	HAC/grc dewatered	37.8	8.9	—
10 years	HAC/grc dewatered	32.2	12.1	10.5
Start	HAC/grc (non-dewatered)	25.3	7.7	—
10 years	HAC/grc (non-dewatered)	29.6	11.7	11.0

7.2 Supersulphated cement (SSC) composites

Supersulphated cement (BS 4248: 1974) made by mixing granulated blast-furnance slag, calcium sulphate and a small quantity of an activator — usually lime or OPC — has been used in the UK as well as in Europe for a number of years. This cement has proved particularly useful in applications where resistance to seawater, sulphates, weak acids and other aggressive agencies is of prime importance.

The properties of SSC may vary widely depending on the nature of the slag and the activator used. At BRE two different commercially blended SSC varieties, Frodingham from UK and Sealithor from Belgium, were used. A quick-setting SSC formulated at BRE by mixing 83% slag, 15% retarded hemihydrate gypsum plaster and 2% hydrated lime were also used in this study. The properties of grc composite boards made from these cements and kept in three different environments for up to ten years are given in Table 7.3. Typical tensile stress-strain curves of five-year old grc made from Frodingham SSC and kept in different environments are shown in Fig. 7.3.

In air the long-term properties of SSC/grc are comparable to those of OPC/grc (see Table 4.1) in terms of strength and impact, but SSC composites have much lower values of LOP and Young's modulus. As these are matrix controlled properties the differences from OPC composites must be ascribed to the matrix phase. In water storage composites made from Frodingham cement have performed appreciably better after ten years than grc made from Sealithor, and the properties of the latter are no better than those of ten-year-old OPC/grc. The ten-year properties of water-stored grc made from Frodingham cement are considerably better than those of OPC/grc in terms of strength and impact resistance, and the composites have retained a large proportion of its very high failure strain after ten years.

In natural weathering, samples of SSC/grc made from all three supersulphated cements have retained a very high proportion of their tensile strain capacity after ten years (Table 7.3). As the ultimate failure strain capacity of fibre cement composites is directly related to their impact resistance it is to be expected that the impact strength of SSC/grc on long-term weathering will be much better than that of their OPC counterpart, and this is borne out by the ten-year results. However, composites made from both Sealithor and quick-setting SSC have given very poor bending and tensile strength on long-term weathering. The Young's modulus and LOP values measured on

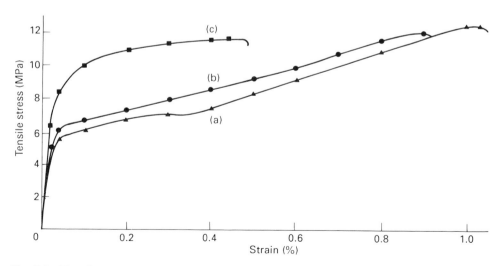

Fig. 7.3 Tensile stress-strain curves for grc made with Frodingham supersulphated cement stored: (a) in air at 65% RH; (b) on the natural weathering site at BRE; (c) under water at 20°C for five years.

Table 7.3 Properties of grc made from supersulphated and Portland blastfurnace slag cement.

| Matrix | Property | 28 days Air (moist cured) | Age and storage conditions |||||||||
| | | | 1 year ||| 5 years ||| 10 years |||
			Air 60–65% RH	Water	Weather	Air 60–65% RH	Water	Weather	Air 60–65% RH	Water	Weather
PBFC	MOR (MPa)	38	33	25	30	27	20	19			
	LOP (MPa)	14	13	13	11	9	14	11			
	UTS (MPa)	14	10	10	9	—	—	—			
	E (GPa)	25	26	32	—	—	—	—			
	IS (kJ/m²)	23	16	9	13	14	4	7			
SSC (Frodingham)	MOR (MPa)	40	35	43	40	31	34	34	32	31	26
	LOP (MPa)	11	10	17	15	9	16	12	11	16	8
	UTS (MPa)	15	13	13	15	13	11	13	12	12	12
	E (GPa)	26	20	35	22	20	32	21	20	35	16
	IS (kJ/m²)	28	27	21	27	23	17	21	22	14	16
	ε_μ (microstrain)	12280	10840	8870	10410	10350	4675	9175	10330	4290	8710
Quicksetting SSC	MOR (MPa)	35	40	40	33	—	29	16	—	—	10
	LOP (MPa)	14	14	17	12	—	17	8	—	—	4
	UTS (MPa)	12	14	13	13	13	11	8	—	—	7
	E (GPa)	30	27	29	30	22	33	13	—	—	11
	IS (kJ/m²)	25	25	20	21	—	7	15	—	—	15
	ε_μ (microstrain)	9770	10700 (40% RH)	7030	10250	11040 (40% RH)	790	8170	—	—	7670
SSC (Sealithor)	MOR (MPa)	26	26	25	25	23	18	20	22	16	16
	LOP (MPa)	12	5	11	10	5	12	5	6	13	5
	UTS (MPa)	10	10	9	10	—	—	8	10	5	7
	E (GPa)	26	16	23	—	—	—	9	14	21	8
	IS (kJ/m²)	18	23	13	17	19	10	13	20	3	11
	ε_μ (microstrain)	10390	10730	7240	7230	—	—	5340	9200	360	6000
									9 years 3 months		
SSC & Sand (70 30) (Sealithor)	MOR (MPa)	29	31	27	24	26	20	21	21	18	18
	LOP (MPa)	12	9	13	12	8	12	6	6	15	7
	UTS (MPa)	10	—	—	—	—	—	—	8	7	7
	E (GPa)	26	—	—	—	—	—	—	16	25	14
	IS (kJ/m²)	19	23	15	17	20	8	13	17	7	10
	ε_μ (microstrain)	10100	—	—	—	—	—	—	11200	515	6400
									9 years 3 months		

these composites are also substantially lower than those of composites made from Frodingham cement. It seems that the matrix phases in these two types of composites exhibiting poor weatherability have been adversely affected.

The superior strength retention of some of the SSC composites compared to their OPC counterparts can be explained partially at least by the greater strength retention of the Cem-FIL fibre reinforcements in the less alkaline SSC as argued by Proctor and Yale [9]. For SSC composites another factor — carbonation — plays a very important role.

The strength development in SSC is due, partly, to the formation of ettringite ($3CaO.Al_2O_3.3CaSO_4.32H_2O$). Atmospheric CO_2 reacts with ettringite, forming $CaCO_3$ and gypsum and a large quantity of water. When the water evaporates, leaving additional porosity in the cement, the matrix dependent properties of the composites assume lower values. Detailed microstructural work[13] has shown that carbonation destroys the ettringite that bonds the unhydrated slag and calcium silicate hydrate regions in SSC, and therefore even the relatively small amount of ettringite present in the cement composites may exert a major influence on their properties. Under wet environments the cement hydrates to a much greater degree, producing a much denser composite which is less permeable to CO_2, and hence its long-term, matrix dependent properties are better than those of the corresponding air-stored values.

The deleterious effect of carbonation on SSC strength and hence on some of the properties of composites made from this cement can be reduced by using these composites underground or under wet conditions. If the surfaces of the Cem-FIL/SSC composites are compacted, for example by casting against steel, or they are provided with a thin protective layer of a sand/cement mortar, the resistance of the material to atmospheric carbonation can be improved substantially. Using a 1–2 mm thick protective layer of white cement on Cem-FIL/SSC composite sheets and subjecting the material (after 28 days' initial cure) to accelerated carbonation, i.e. CO_2 gas under 0.7 MPa pressure for seven days, it was found that the MOR, LOP and IS of the material were 24 MPa, 13 MPa and 17 kJ/m^2, respectively, compared to 21 MPa, 11 MPa and 13 kJ/m^2 for the uncarbonated specimens. When accelerated carbonation was carried out at ambient pressure and temperature at 50% RH, the MOR of the protected composite remained the same after 90 days as that of the starting material (21 MPa), but its impact strength increased from 13 to 21 kJ/m^2 and LOP decreased from 11 to 8 MPa. These results can be explained in terms of limited carbonation of SSC matrix. The long-term properties of grc composites made from SSC and provided with protective render are given in Table 7.4.

Proctor et al.[12] have also studied the long-term properties of grc made

Table 7.4 Properties of SSC composites with protective renders.

Property		28 days	90 days		5 years	
			Air 40% RH	Weather	Air 40% RH	Weather
(A) SSC/sand protective layer						
MOR	(MPa)	33	34	32	25	24
LOP	(MPa)	9	9	15	5	9
IS	(kJ/m^2)	19	23	18	18	12
(B) OPC/sand protective layer						
MOR	(MPa)	28	28	26	23	23
LOP	(MPa)	16	10	14	7	12
IS	(kJ/m^2)	19	20	18	16	10

from SSC and Cem-FIL and Cem-FIL 2 AR glass fibres. They confirm the inferior properties of these composites after long-term natural weathering compared to storage under water. In particular, LOP and Young's modulus, which are matrix controlled properties, have shown very substantial reduction after natural weathering for grc made from both fibres. In long-term water storage the Cem-FIL 2/composites have retained most of their initial strength and toughness (see Table 7.5). For composites made from original Cem-FIL fibres substantial reductions in strength and toughness were observed in water storage irrespective of the type of SSC used. In this respect their ten-year results for Cem-FIL/grc made from Frodingham SSC are far more pessimistic than those obtained at BRE. The reasons for this discrepancy are not clear at the present time but it raises questions about the long-term potential of Cem-FIL/grc made from supersulphated cements. Standard concrete cubes made from the batch of Frodingham cement used in the BRE work have given acceptable compressive strength (> 60 MPa) when stored under water for ten years but very poor results when kept in air.

Table 7.5 Properties of spray-dewatered SSC/grc made with 5.2% Cem-FIL 2 after storage in water at 20°C (Reference 12).

Test time	MOR (MPa)	LOP (MPa)	Modulus (GPa)	Impact strength (kJ/m^2)	Bending strain (%)
Start	36.1	11.9	15.0	21.4	1.21
10 years	39.1	13.0	15.5	17.3	1.07

7.3 Portland blastfurnace cement (PBFC) composites

The properties of the Cem-FIL/PBFC under three different storage conditions are listed in Table 7.3. The bending and impact strength of these composites up to five years are very similar to those of composites made from OPC. Since the OPC content of the batch of PBFC used in the present work was of the order of 80% by weight, this similarity in the results is to be expected.

7.4 Other cement composites

From time to time cements are developed for special purposes. One such cement, regulated set cement, was developed in the USA in the 1960s[14] and found limited use in the construction industry for its rapid set and strength development. Grc sheets made from such a cement and reinforced by AR glass fibres have been in commercial production and use in Europe but the properties of these composites have not been reported in the open literature.

Very recently the Chichibu cement company in Japan has developed a special cement, containing calcium aluminate sulphate ($3CaO.3Al_2O_3.CaSO_4$) as a major constituent, for use in grc composites. A similar cement is also used in China[15] in the grc industry. For reinforcing such sulphoaluminate cements in Japan as well as in China, continuous AR glass fibres similar to Cem-FIL are being used. Hyashi et al.[16] have recently described the properties of such composites. The accelerated ageing tests have shown that AR glass fibres are more durable in the special cement than in OPC. There was little or no loss in the bending strength of the composite made from the sulphoaluminate cement after 56 days in water at 70°C, and the failure strain also remained high after such a treatment. After 28 days in water at 70°C, the UTS of the composite remained unchanged but the tensile failure strain showed a small decrease. The impact strength of the composite was unaffected after 28 days in water at 70°C. The drying shrinkage of the composite at 20°C and 65% RH was about one-fifth of grc made from OPC.

From the studies made so far it can be concluded that the long-term values of some of the strength properties of cement composites made from Cem-FIL AR glass fibres and HAC or SSC under continuously wet or UK natural weathering conditions are better than those obtained with their OPC counterparts. It must be remembered, however, that, as a matrix for reinforcement by fibres, HAC, and to a lesser extent SSC, have limitations showing significant reductions in strength with time – HAC because of the well-known 'conversion' reactions, and SSC due to atmospheric carbonation.

Composites made from PBFC with relatively small proportions of slag have not shown any advantages over those made from OPC. The results obtained in Japan and China on grc made from sulphoaluminate cements are very encouraging but it will perhaps be several years before one could assess the long-term prospects for such composites. It should be noted that the long-term strength properties of OPC composites are expected to be vastly superior when Cem-FIL 2 fibres are used as the reinforcement. Also, it is important to remember that most special cements are more expensive than OPC. A great deal of attention, therefore, needs to be paid before non-OPC cements are selected for manufacturing grc. Nevertheless, in special situations they may have a role to play.

There is a current body of opinion that links the long-term durability of grc with the absence of crystalline $Ca(OH)_2$ in the cement hydration products. Nearly all non-Portland cements produce little or no $Ca(OH)_2$ on hydration, and the superior performance of these cements as the matrix material for grc when compared with OPC has been ascribed by some to this particular feature. This point will be discussed further in a later chapter.

7.5 Glass fibre reinforced autoclaved calcium silicate (grcs)

Autoclaved calcium silicate material in the form of sand—lime bricks has been used in the construction industry for many years. Until very recently and before the health hazards of asbestos were widely recognised, certain types of insulation boards containing asbestos were produced in many countries by autoclaving a mixture of lime and silica to which a suitable amount of Portland cement was often added. The hydrated calcium silicate matrix that is formed is of low density, and such boards having nominal densities of 576 and 720 kg/m^3 found extensive use in the building industry, one particular variety being almost exclusively used until recently in the partition systems in ships.

Inorganic glass fibres such as rock wool are corroded very severely in the autoclave in the environment of a hydrating lime/silica mix. With the advent of alkali-resistant glass fibres such as Cem-FIL, attempts were made[17-19] to replace asbestos by such fibres in autoclaved calcium silicate insulation boards.

In the material developed at BRE[17, 18] the matrix formed by autoclaving is similar to that of the traditional product, but asbestos fibres are replaced by Cem-FIL alkali-resistant glass fibres. Before autoclaving, the boards were first made by the BRE spray-dewatering method and partially dried. Some of the fabrication parameters were varied: CaO/SiO_2 molar ratio 0.8—1.2, autoclave temperature 150—180°C, autoclave time 2—21 hours and glass content 1.2–6.5 volume %, the latter being a major factor in controlling strength. The best results for a glass content of 2.8 volume % were obtained with boards produced from hydrated lime and kieselguhr (a source of silica) in a CaO/SiO_2 molar ratio of 0.8, with glass fibres cut to 32 mm lengths from continuous strands, and by autoclaving for 20 h at 180°C.

The principal phase in the BRE material (grcs) is the well-crystallised 11.3Å (1.13 nm) tobermorite ($5CaO.6SiO_2.5H_2O$). Portlandite ($Ca(OH)_2$) is virtually absent. Tensile strength tests on glass fibres extracted from the autoclaved boards showed a reduction of fibre strength of more than 50% from the initial strength in the strand, but the strength is not expected to change significantly with time after manufacture of the board. Specifically, no long-term loss in strength should occur in indoor applications where grcs is most likely to be used.

At similar densities, most mechanical properties of grcs are similar to or better than those of their asbestos counterparts (Table 7.6). The asbestos products have, however, better resistance to screw-pulling.

Although grcs was developed as a shipboard, its use may extend to other areas where good insulation and fire protection are needed and where resistance to damage by water is desirable. The material has successfully withstood the appropriate small-scale fire tests carried out in accordance with BS 476, Part 8.

When grcs is cut, the glass fibres released in the dust are too large to be respirable. In one series of experiments[20] the concentration of fibres produced from asbestos boards, while sawing with typical shipyard joinery tools, was 2000 times greater than that derived from grcs. The median values of length and diameter of the fibres in the dust were approximately 100 µm and 12.5 µm for the glass fibres, whereas the corresponding values with asbestos

Table 7.6 A comparison of grcs boards with special purpose asbestos boards.

Fibrous reinforcement	Cem-FIL glass fibre			Asbestos	
Cem-FIL, 32 mm (vol %)	2.8		3.4		
Age (years)	0.01	9*	0.01	As purchased	
Thickness (mm)	8	8	15	12	9
Density (kg/m^3)	540	560	610	570	690
Bending strength (MPa)	8	6.5	7.5	6	6.5
Impact strength – Izod (kJ/m^2)	8.5	10.5	7	4.5	2.5
Screwholding no. 8 into face (N)$^+$	200	—	460	470	490

* Kept in air at 40% RH and 20°C.
$^+$ Value increases with thickness.

were less than 5 μm and 1 μm, respectively. Over 98% of the fibres sampled while the asbestos board was machined had diameters below 3–4 μm, indicating that they were almost all respirable whereas none of the glass fibres sampled were small enough to be respirable.

The results described above are promising enough to suggest that if Cem-FIL fibres are replaced by fibres such as Cem-FIL 2 or Asahi Super and the matrix is suitably formulated by using the relevant experience of the industry, it may be possible to produce a much stronger grcs material.

Chapter Eight
Microstructure of Grc and Glass/Cement Bond

The microstructure of pastes made from Portland and other hydraulic cements evolves gradually with time as the hydration proceeds. In recent years attention has been paid to the evolution of microstructure during the hydration of Portland cement in order to explain the mechanisms of such hydration and the effect it might have on the development of properties such as setting and hardening, permeability and strength. When fibres or other materials such as aggregates are added to the paste, it is assumed that the hydration sequences for cement are not altered in any major way but obviously the microstructure development of the composite is affected. In this chapter we discuss the important microstructural features of grc, how they change with time in different environments and the effects these changes may have on the properties of grc, including its durability. The exact method of manufacturing grc (i.e. by premixing or spraying or filament winding) will have an important bearing on the microstructure of the composite, but here we shall be concerned mainly with grc made by the spray-up method where an approximately 2−D random dispersion of short fibres in the composite is obtained. The microstructure of grc described in this chapter mainly refers to grc made from Cem-FIL AR glass fibres.

In Chapter 2 we have discussed the importance of the fibre/cement bond in the development of the mechanical properties of brittle-matrix composites. Here we describe the methods that are used for measuring the strength of the bond between glass fibres and cement.

8.1 Initial microstructure

Although light optical microscopy has been used in the examination of the microstructure of grc in some cases[1], most investigators have preferred to use the scanning electron microscope (SEM) for obtaining detailed information. Most commonly, rough fracture surfaces of the specimen have been studied with the SEM, and only recently it has been possible to examine the relationship between microstructure and crack path in grc by stressing the composite *in situ* in the SEM specimen chamber[2].

The fracture surfaces of 'standard' grc made from ordinary Portland cement and stored in dry (40% RH) and moist (90% RH) air for 90 days are illustrated in Figs 8.1 and 8.2. Several features of the surfaces are immediately apparent. Firstly, the fibre strands are essentially integral in all samples and can therefore be considered as the reinforcing element (it is possible, however, to obtain a greater degree of filamentisation of the strand in grc by choosing a suitable size for the fibre). Secondly, there is evidence of considerable fibre pull-out from the matrix. Grc specimens naturally weathered for one year also show these features (Fig. 8.3). The fracture surfaces of grc stored in a relatively dry environment are characterised by the large amount of subsidiary cracking which occurs during failure, resulting in a large fracture surface area.

It is widely recognised that pull-out of fibres is one of the main mechanisms by which brittle-matrix composites such as grc gains toughness. This is particularly true for 'young' grc. The degree of fibre pull-out is controlled by the strength of the interfacial bond and it is instructive to examine the microstructure of the glass/cement interface. In Figs 8.4(a), 8.4(b) and 8.5 SEM photographs of the fibre/matrix interface in grc stored in dry (40% RH) and moist (90% RH) air for 90 days are shown. Fibre pull-out grooves are clearly visible but in the drier specimen there is evidently more porosity and the fibre/matrix interface is less dense. In 90-day dry air-stored grc the fibres within a bundle remain quite 'loose'; the cement hydration products have not at this stage penetrated into the bundles in large quantities. In all environments normal hydration products of Portland cement, $C-S-H$ and $Ca(OH)_2$ are present, the latter appearing as large tablets in pore spaces. Since the fibre strand, which is the reinforcing element, is porous, growth of large $Ca(OH)_2$ crystals in and around the glass filaments in the strand is an important feature of grc microstructure. These photographs are taken from the work of Stucke and Majumdar[3], who suggested that the whisker-like hydration products seen in Fig. 8.4(b) could be $C-S-H$, and obtained evidence that indicates that these whiskers bond well to the glass fibres. Direct experimental evidence[4] suggests, however, that bond strength between cement and glass fibre strands is low, of the order of $1-3$ MPa.

8.2 Microstructure changes with time

The microstructure of grc changes with time in a significant way when the composite is kept continuously in a wet environment. In Figs 8.6(a) and (b) fracture surfaces of grc stored for five years in water at 20°C are shown.

Fig. 8.1 Fracture surfaces of grc stored for 90 days in dry air (40% RH) at 20°C.

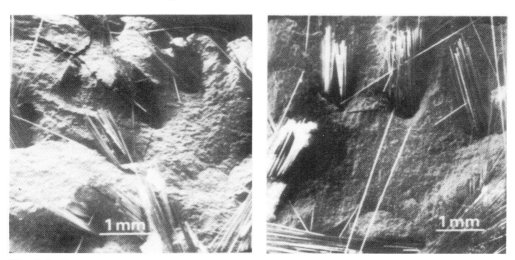

Fig. 8.2 Fracture surfaces of grc stored for 90 days in moist air (90% RH) at 20°C.

Fig. 8.3 (left) Fracture surface of grc naturally weathered for one year.

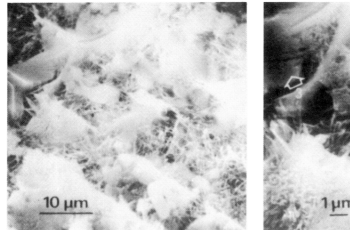
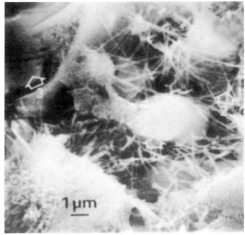

Fig. 8.4 Pull-out grooves showing fibre matrix interfaces in 90 day dry air stored grc. (1) Partially hydrated cement grain; (2) 'whisker'-like hydration products; (3) Ca(OH)$_2$ crystal.

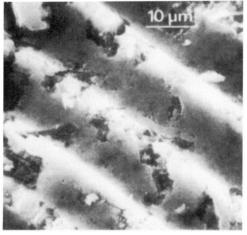

Fig. 8.5 (left) Fibre−matrix interfaces in 90 day moist air stored grc.

Fig. 8.6 (below) (a), (b) Fracture surface of grc stored for five years in water at 20°C.

Fig. 8.7 is a similar SEM photograph from a grc sample naturally weathered for five years. It is evident that the fracture surfaces in the photographs in Fig. 8.6 are almost planar with little or no evidence of fibre pull-out, most fibres failing at one or the other of the fracture faces. By contrast the amount of pull-out has not decreased in dry air-stored specimens after five years (Figs 8.8(a) and (b)) from that seen after 90 days (Fig. 8.1), and even after 20 years the fracture surface of such specimens remains fibrous (Fig. 8.8(c)). The microstructure of grc naturally weathered for five years (Fig. 8.8) is closer to that of the wet-stored material, although in this case there is evidence for some fibre pull-out. Fig. 8.9 illustrates interfaces in five-year dry air-stored grc, and it can be seen that they are very similar to those in the 90-day material. In the five-year wet-stored material (Fig. 8.10(a)) the interfaces become almost fully dense, reflecting the much larger volume of hydration products formed from the cement in the wet environment. Most of the material at the interface is[1,3] $Ca(OH)_2$ and some of the whisker-like crystals are observed in the residual porosity. Fig. 8.10(b) shows layers of $Ca(OH)_2$ covering glass fibres in 20 year old water-stored grc. The microstructure of the interface in the five-year naturally weathered material (Fig. 8.11(a)) varies from the porous type observed in dry storage to the fully dense of the wet-stored material, different types of interface being observed in different parts of the sample. The reasons for this are not clear but it may be due to the variation in hydration conditions in different parts of the composite, resulting from changes in atmospheric humidity and temperature experienced during natural weathering. The microstructure of grc after 17 years of natural weathering at Garston is illustrated in Fig. 8.11(b). A nearly planar fracture face is revealed.

The role of $Ca(OH)_2$ crystals in grc microstructure is considered by many to be very important. In particular, it has been pointed out[3,5] that densification of the fibre/cement interface caused by the precipitation and growth of $Ca(OH)_2$ crystals in and around glass fibre bundles may be linked with the embrittlement of grc in wet conditions. In steel fibre reinforced mortar specimens Pinchin and Tabor[6] found a $Ca(OH)_2$ rich layer within about 10 μm of the wire. The affinity of $Ca(OH)_2$ crystals for glass is also demonstrated in studies by Diamond et al.[7] and Mills[8]. In grc Ca^{++} in the pore solution has easy access to the interior of the fibre bundle, which is porous. Precipitation of $Ca(OH)_2$ crystals in the inter-filament space will take place in due course, and in this respect it is possible that the glass fibre surface may provide suitable nucleation sites.

Fig. 8.7 Fracture surface of grc naturally weathered for five years.

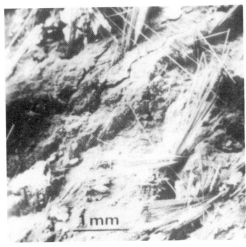

Fig. 8.8 Fracture surfaces of grc stored for five years (a), (b), and for 20 years (c) in dry air (40% RH) at 20°C.

(a) (left)
(b) (below left)   (c) (below right)

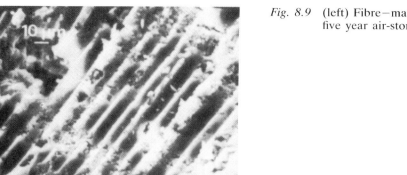

Fig. 8.9 (left) Fibre−matrix interfaces in five year air-stored grc.

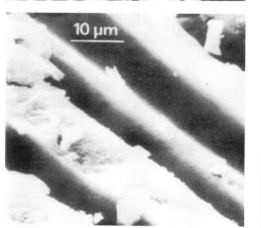

Fig. 8.10 (a) (left) Fibre−matrix interfaces in five year water-stored grc; (b) (right) Ca(OH)$_2$ deposits surrounding glass fibres in grc after 20 years in water at 20°C.

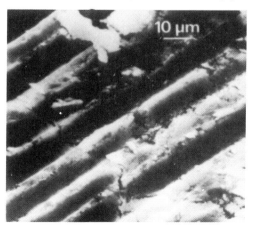
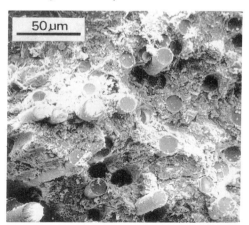

Fig. 8.11 (a) (left) Fibre−matrix interfaces in five year naturally weathered grc; (b) (right) microstructure of grc after 17 years of natural weathering at Garston.

8.3 Microstructure of Cem-FIL 2/grc

It has been discussed in preceding chapters that when Cem-FIL 2 AR glass fibres are used in reinforcing cement instead of the original Cem-FIL fibres, the resultant grc composite retains a much higher proportion of their initial strength and toughness in the longer term than Cem-FIL/grc, even in a continuously wet environment. It has been demonstrated that the microstructure of wet-stored Cem-FIL 2/grc is less dense near the reinforcement, and solidification of the fibre strand does not take place in this case as with Cem-F!L 2/grc[5]. The microstructure of Cem-FIL 2/grc made from neat OPC and weathered at BRE for about four and nine years shown in Figs. 8.12(a) and 8.12(b), respectively, is not very different from that of air-stored grc made from Cem-FIL fibres (Fig. 8.1). The fibre strand has remained integral after the grc specimen has failed. A close examination of the microstructure of Cem-FIL 2/grc specimens kept in different environments for four years or more suggests that the fibre bundle has remained tightly packed (Fig. 8.13(a)) with little space between filaments. But the deposition of cement hydration products including $Ca(OH)_2$ inside the fibre bundle is still not a very rare occurrence (Fig. 8.13(b)), the process being aided by the local 'looseness' in the fibre bundle. It has been pointed out[9] that the special surface treatment provided for Cem-FIL 2 interferes with the normal hydration of cement in and around the fibre bundles, thereby possibly reducing the amount of $Ca(OH)_2$ that can be formed, but this point needs further investigation.

8.4 Microstructure of grc made from non-Portland and blended cements

It has been discussed in preceding chapters that the long-term strength retention in grc (made from Cem-FIL AR fibres) in wet environments is significantly improved if Portland cement is replaced by other cements such as high alumina or supersulphated cement, which is less alkaline. If OPC blends containing large proportions of ground granulated blastfurnace slag or pulverised fuel ash, preferably the former, are used as the binder, the long-term durability of grc is improved. Microstructural studies of grc made from non-Portland cement have not been as detailed as those carried out on OPC/grc; nevertheless some information is available in the literature. The microstructure in these composites differs from that of OPC/grc in several respects: (1) the hydration products are different, e.g. in HAC/grc calcium aluminate hydrates, in SSC/grc abundant calcium aluminosulphate hydrate (ettringite), (2) very little if any $Ca(OH)_2$ in HAC/grc, small amounts in SSC/grc and (3) the void/solid ratio in and around the fibre reinforcement

Microstructure of the Grc and Glass/Cement Bond 151

Fig. 8.12 Cem-FIL 2 fibres in (a) four year and (b) nine year naturally weathered grc.

Fig. 8.13 (a) Cem-FIL 2 fibre strand integrity in weathered grc. (b) Cement hydration products between Cem-FIL 2 filaments in weathered grc.

remains high even in water-stored materials. The resultant effect is that, even in wet environments after many years, fibre pull-out remains as an important mechanism in composite fracture processes. This is evident in the microstructure of very old HAC/grc samples after weathering (Fig. 8.14(a)) or storage in water (Fig. 8.14(b)).

The microstructure of naturally weathered grc composites made from SSC shows some interesting features. It is well known that SSC is easily carbonated on exposure to atmospheric CO_2 and that a friable 'soft skin' is formed. In their study on SSC/grc, Stucke and Majumdar[10] found two types of deposit on grc samples exposed to natural weather:

(a) A powdery deposit, which is whitish in appearance, is easily brushed off the surface. An SEM photograph of this material is shown in Fig. 8.15. It is composed of a mat of randomly oriented columnar crystals of ettringite loosely bonded to the composite specimen surface. The 'soft skin' formulation mentioned in the literature corresponds most probably to the microstructure shown in Fig. 8.15.
(b) A cluster of rhombic crystals, probably of calcite shown in Fig. 8.16 produced by the carbonation of ettringite.

The microstructure of the fracture surfaces of SSC/grc kept in relatively dry air, water and natural weather for up to two years is very similar and roughly corresponds to the microstructure of the air-stored OPC/grc (Figs 8.1 and 8.2) in essential features such as extensive fibre pull-out and subsidiary cracking. Weathered specimens of SSC/grc show evidence of disintegration of the fibre/matrix interface by carbonation (Fig. 8.17). It is thought that the ettringite crystals bonding the areas of ettringite and C–S–H in contact with the fibre are destroyed by the action of CO_2, leading to increased porosity and easier further carbonation. The destruction of the dense SSC matrix and the glass fibre/matrix bond results in a decrease in the limit of proportionality (LOP) values of the composite[10].

In blended cements, a part of Portland cement is replaced by other materials, notably pulverised fuel ash (pfa), ground granulated blastfurnace slag (ggbs) and silica fume. The industrial importance of blended cement is increasing rapidly. Since pfa and silica fume are efficient pozzolanas, and ggbs latently so, in grc made from blended cements these materials react with $Ca(OH)_2$ and other alkaline hydroxides derived from OPC. The amounts of C–S–H become proportionately higher. The microstructure in these composites is denser than the corresponding OPC/grc, and it produces lower permeability.

The microstructures of grc made from blended cements and kept in different environments have been examined by SEM. In general they show

Microstructure of the Grc and Glass/Cement Bond 153

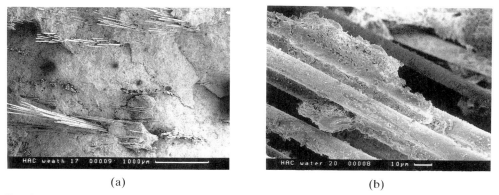

Fig. 8.14 Microstructure of grc made from HAC: (a) after 17 years of weathering at Garston, (b) after 20 years in water at 20°C.

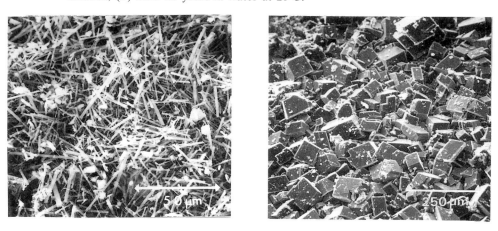

Fig. 8.15 Powdery deposits consisting mainly of columnar crystals of ettringite.

Fig. 8.16 Crystalline deposits of calcite.

Fig. 8.17 Microstructure of the SSC matrix after carbonation: (a) one year old grc; (b) two year old grc after weathering.

the same features as illustrated by the SEM photos of OPC/grc described previously. In some cases, in wet or natural weathering conditions of storage these grcs show evidence of appreciable fibre pull-out instead of fibre fracture, as is the case with ordinary OPC/grc. In Fig. 8.18 the microstructure of the fracture surface of grc made from a commercially blended OPC/pfa mixture and containing the original Cem-FIL fibres is shown. The material was kept under water for five years.

The other important aspect of the microstructure of grc made from blended cements is the extent to which $Ca(OH)_2$ crystals grow in and around the fibre bundles in these materials. There is some evidence to suggest that when OPC/pfa blends are used as the matrix instead of the neat cement, the size of the $Ca(OH)_2$ crystals is much smaller and these do not fill up the void space completely inside the fibre bundle in a way that is characteristic of OPC/grc containing Cem-FIL fibres. Consequently, there is still the possibility of slippage of glass filaments inside fibre bundles when the composite is stressed. The pull-out grooves seen in many SEM photographs of these samples may be evidence for such a proposition. For the same type of curing an OPC/grc sample would give a planar surface microstructure. Figs. 8.19(a) and 8.19(b) show the microstructure of grc made from 60% OPC, 40% pfa and Cem-FIL fibres after 17 years of natural weathering and water storage. The evidence for abundant fibre breakage is very clear, but less so of fibre pull-out.

In Figs. 8.20 and 8.21 the microstructures of Cem-FIL 2 glass fibre bundles in grc composites made from blended cements are shown. By and large the microstructures seen are similar to those noted for the corresponding composite made from neat cement or cement plus sand (Fig. 8.12).

Fig. 8.22 shows that after nine years of natural weathering the fracture surface of grc composites made from Cem-FIL 2 and containing ggbs remains essentially fibrous and this accounts for the high impact resistance of such composites (Table 5.6). A very similar microstructure is displayed by composites containing pfa after long-term weathering. Fig. 8.23 illustrates the point that penetration of fibre bundles by cement hydration products is not uncommon in old grc samples made from blended cements and Cem-FIL 2 fibres.

8.5 Microstructure and bond

From the micrographs described earlier in the chapter it is clear that the fibre/cement interface in air-stored grc made from OPC and Cem-FIL AR glass fibres remains porous for a very long time, whereas when the com-

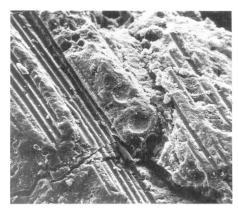

Fig. 8.18 (left) Microstructure of grc made from Pozament.

(a)

(b)

Fig. 8.19 Microstructure of grc containing pfa: (a) after 17 years of weathering at Garston, (b) after 17 years in water at 20°C.

Fig. 8.20 Cem-FIL 2 fibres in grc containing granulated slag after four years' weathering.

Fig. 8.21 Cem-FIL 2 fibres in grc containing pfa after five years' weathering.

156 *Glass Fibre Reinforced Cement*

Fig. 8.22 Microstructure of Cem-FIL 2/grc containing ggbs after nine years of natural weathering at Garston.

Fig. 8.23 Microstructure of Cem-FIL 2 fibre strand in grc containing pfa after nine years of natural weathering at Garston.

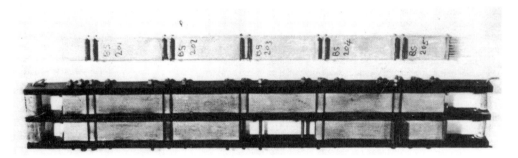

Fig. 8.24 The mould, partly assembled, and a strip specimen.

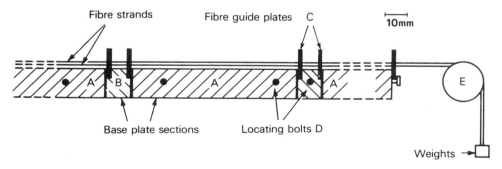

Fig. 8.25 The mould (section view) showing details of the base plate, fibre guides, etc.

posite is stored in a wet environment progressive hydration in and around fibre bundles results in a considerable densification of the interface. One major effect of this densification of the interface, it has been argued[3,5], is the progressive embrittlement of grc in wet and natural weathering conditions. When Cem-FIL 2 fibres are used as reinforcement, even under wet conditions of use lasting a few years the interface still remains porous and fibre pull-out is observed although it is not possible to say how long this failure mechanism will remain unaltered. It has been assumed that the change in microstructure from dry to wet grc produces an increase in the strengths of bond between the fibre and the matrix. The precipitation of $Ca(OH)_2$ in and around the fibre bundles is supposed to be responsible for the increase in bond strength.

8.5.1 Bond strength measurement

Two alternative approaches have emerged for the determination of the strength of the interfacial bond in fibre cement composites. Bond strength can be obtained from measurements of crack spacings, and Aveston et al.[11] have used this method for aligned continuous fibre composites. Oakley and Proctor[12] have also applied the method, this time to grc (random 2–D) sheets. In grc the most frequently used bond strength value is that obtained from a pull-out test in which the load to debond and pull out a fibre from a button of the matrix is measured[13,14]. This apparently simple test presents considerable difficulties even when single rods or filaments are tested, both because of the high variability of the results and because of problems in interpretation. Uncertainty also arises from the multi-filament nature of the fibre strand used in commercial grc composites. A multi-fibre pull-out test is described below, which was developed at BRE[4] to reduce the variability of the results, and at the same time to simulate more accurately conditions in an actual composite.

8.5.1.1 THE MULTIPLE FIBRE PULL-OUT TEST

In the test, 16 parallel fibre strands mounted in two rows of eight are loaded simultaneously. The test specimens are cut from two strips (each providing five pull-out specimens) produced using a specially designed mould (Figs 8.24 and 8.25). The mould consists of three bars which form the 'sides' and central separator of the assembled double mould. These are separated at each end by circular cylindrical rollers (E, Fig. 8.25) over which the ends of the fibre strands are located. The 'base plate' consists of sections (types A and B) shaped to locate the brass guide plates C, through which the strands are threaded. The sections are held in the mould by long brass bolts D which are fitted from one side of the mould, through the base plate sections, etc. to the other side. The brass guide plates contain two rows of eight holes and are fitted in pairs as shown. They perform the double function of holding the fibres in position during specimen fabrication and of transferring the load to the finished specimens during the test. The guide plate adjacent to the pull-out section of the finished specimen acts also as an artificial crack. The holes in the guide plate are hemispherically bevelled on the side facing the partner guide plate to encourage matrix friction and pull-out at a well-defined point. The mould has the advantage over earlier designs in that, being fully demountable, it can be removed with minimum damage to strip specimens. To further facilitate this release, the assembled mould and guide plates are coated with a PTFE release agent before the fibres are threaded.

Sixteen fibre strands are threaded by hand through the guides in the assembled moulds and are taped over the rollers E at one end. Small weights (steel nuts) are tied to their free ends to hold them taut while the matrix is poured. The cement paste is hand mixed and, to ensure even filling of the mould and penetration of the paste between the strands, the mould is vibrated on a vibrating table while the paste is being poured. The specimens are allowed to set and are stored as strips. Each strip is subsequently cut into five specimens leaving a length of several centimetres below one of the brass guides (the lower one in the test machine) to anchor the fibres at this end. The pull-out length is the length allowed above the other (top) brass guide. The 'average bond strength' is obtained from the maximum load achieved as the fibre strands are withdrawn from the top (pull-out) length in a direct tensile test.

At sufficiently short embedment lengths most of the fibre strands remain intact, and the average bond strength τ is simply given by

$$\tau = \frac{P}{np\ell} \tag{8.1}$$

where P is the maximum load achieved, n is the number of fibre strands, p is the 'fibre' perimeter at the interface, and ℓ is the embedment length. The product of perimeter and bond strength ($p\tau$), that is the pull-out force per fibre strand per unit length, is calculated rather than τ alone, because of the uncertainty of the value of p that applies to the multi-filament strands of variable shape in a porous matrix. In effect $p\tau$ is the slope of the initial part of the load (per fibre strand)/embedment length curve.

8.5.2 Glass/cement bond

Cem-FIL and Cem-FIL 2 fibres used by Laws et al.[4] in their study of the glass/cement bond were in the form of strands consisting of glass filaments approximately 12.5 μm diameter. Pull-out specimens were made using ordinary Portland cement and a water/cement ratio of 0.3.

During the seven-day initial cure period of the Cem-FIL 2 fibre samples at 90% RH and 20°C, the value of $p\tau$ increased from 1.4 N/mm at one day to 4 N/mm after seven days. On transfer to water storage at both 20°C and 50°C, $p\tau$ decreased. The effect of storage time and condition is shown in Figs 8.26 and 8.27. For water storage at 20°C, $p\tau$ increased with storage time. Storage in air at 40% RH and 20°C led to lower values of $p\tau$, and storage time had little effect (Fig. 8.26). At 50°C $p\tau$ for Cem-FIL 2 decreased; for Cem-FIL, $p\tau$ appeared to reach a maximum value and then decrease

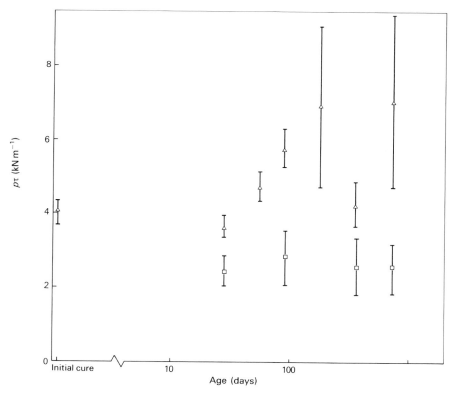

Fig. 8.26 Variation of bond strength with time of storage at 20°C. Cem-FIL 2 fibre, △ cured in water; □ cured in air at 40% RH. p is the perimeter of the fibre strand. The 90% confidence limits are shown.

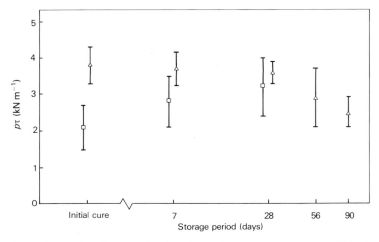

Fig. 8.27 Variation of bond strength with time of storage in water at 50°C. △ Cem-FIL 2 fibre; □ Cem-FIL fibre. The 90% confidence limits are shown.

slowly (Fig. 8.27). Prolonged water storage at 50°C of Cem-FIL 2 fibre samples led to lower values of $p\tau$ than water storage at 20°C.

In the tests of Cem-FIL fibre initial curing was followed by storage in water at 50°C only, for a maximum of 28 days. Nevertheless where there is a comparison, values of $p\tau$ for Cem-FIL 2 fibre are generally higher than those for Cem-FIL fibre (Fig. 8.27). Unfortunately the extent of fibre breakage made it impossible to obtain values of $p\tau$ for the Cem-FIL fibre specimens in water at 50°C at storage times beyond 28 days, and it is not known therefore whether the trend of increasing bond strength with time continues.

In tests of fibres with long embedment lengths, most fibres broke and the maximum load achieved approached the failure load of the array of fibre strands. Provided that all the strands broke, the strand strength could be calculated. However, in some cases, particularly in the early stages of storage, some did not break, and the average maximum stress in the fibre strands during 'pull-out' of the 10 mm embedment length then underestimated the strength of the fibre strand. Nevertheless the results suggested that the strand strength of the Cem-FIL 2 fibre stored in water at 20°C did not change much over the period of the test. Storage in water at 50°C resulted in decreased strength for both Cem-FIL fibre and Cem-FIL 2 fibre, the latter consistently showing the higher strength.

Oakley and Proctor[12] calculated bond strengths from crack spacings for grc sheets made by the spray-suction method and stored in air. The values they deduced were of the order of 1 MPa or less, and these decreased with time. Assuming a value of 2.8 mm for the perimeter of the glass fibre

(Cem-FIL AR) used, the authors deduced average values of $p\tau$ of approximately 3 Nmm^{-1}, which is in agreement with the values reported later by Laws et al.[4] Using G20 glass pencils or rods of ~1 mm diameter, de Vekey and Majumdar[14] had reported bond strength values of the order of ~20 MPa between glass and cement but it appears that a much weaker bond obtains in grc.

8.6 Microstructure and cracking

The details of cracking in grc are revealed in microstructural studies of fracture surface produced in bending or tensile tests. The fracture surfaces of air-stored grc (Fig. 8.8) are characterised by the large amount of subsidiary cracking which occurs during failure, resulting in a large fracture surface area. The amount of subsidiary cracking decreases with storage time in the wet-stored material (Fig. 8.6). The primary crack which is responsible for the ultimate failure of the composite may result from the intersection of these subsidiary cracks of which there are three types: (a) multiple transverse cracks, which run nearly parallel to the primary crack surface, i.e. perpendicular to the applied tensile stress; (b) matrix shear cracks, running parallel to the applied stress and the plane of the array of glass fibres; (c) interfacial cracks observed around each fibre bundle, resulting from the failure of fibre-matrix bond and pull-out of the fibres. These cracks are often linked by matrix shear cracks and produce delamination of the composite (Fig. 8.28). In 'young' grc some fragmentation or crumbling of the matrix around the fibre strands has been observed (Fig. 8.29), resulting in the removal of a wedge-shaped section of the matrix.

Stucke and Majumdar[3] observed that nearly all fibre failures which occur, without pull-out, in the young and old grc made from Cem-FIL fibres are at the crack surfaces. They suggested that this could be due to fibre failure by bending at the crack surface. A few fibres fail beneath the crack surface creating holes (Fig. 8.30). These probably fail as a result of uniaxial tensile stress in the fibre. Premature failure of fibres suggests that these are damaged.

The fracture surfaces of old grc kept in water or natural weathering environment (Figs 8.6 and 8.7) show little or no evidence of subsidiary cracking, although some limited shear failure of the matrix around the fibre bundles is usually observed.

Bentur and Diamond[2] have recently studied the fracture patterns produced when a crack advancing from a notch in cement paste intersected a Cem-FIL 2 alkali-resistant glass fibre strand placed perpendicular to it. The specimens were small notched compact tension specimens that could be wedge loaded in the SEM chamber using a wet cell. Four distinct crack

Fig. 8.28 Delamination in grc stored for 90 days in dry air resulting from the linking of interfacial cracks with matrix shear cracks.

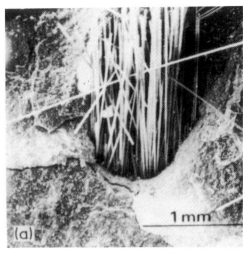

Fig. 8.29 Matrix crumbling around fibre strand in young air-stored grc.

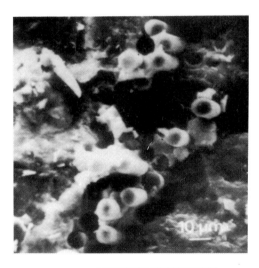

Fig. 8.30 (left) Failure of fibres at the crack surface due to bending stresses in fibres, and below the surface due to uniaxial tensile stress in fibres.

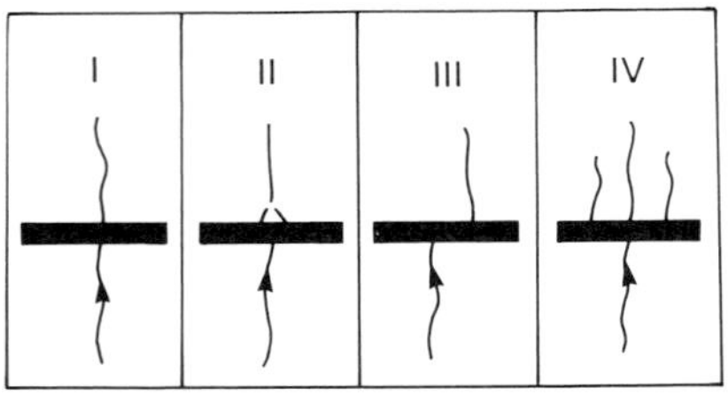

Fig. 8.31 (left) The four cracking patterns observed at the intersection of the propagating crack and the glass fibre strand (Reference 2).

patterns were identified as the load was increased. These are shown schematically in Fig. 8.31. In most cases the fibre caused a shift in the crack path (Type III), and in some specimens microcracking and separation of the crack into two or four branches was observed. The glass fibres maintained their continuity and bridged over the crack. These grc composites were very young and strongly pseudoductile, and the crack patterns observed were very similar to that described by Stucke and Majumdar[3] for air-stored grc. There is some suggestion in the work of Bentur and Diamond on E-glass/cement composites aged by accelerated curing at elevated temperatures that when grc becomes brittle, the crack path is straight and many filaments are broken. This will produce the microstructure seen in Figs 8.6 and 8.7 for water-stored and weathered grc made from Cem-FIL fibres.

Chapter Nine
Durability

The assessment of durability of a new material such as grc poses very difficult problems. Like many other construction materials grc is in use all over the world, both inside and outside buildings and in special applications such as sewer pipes where a permanent, wet chemical environment prevails. The durability of the material will be distinctly different in these different environments and needs to be assessed for each application. In this respect grc is no different from other cement based materials, for example concrete. Many agents, both natural and man-made, have a deleterious effect on concrete. Moisture movements, freeze—thaw and attack by chemicals and carbonation (in the case of reinforced concrete) are only a few examples that receive routine attention from engineers and materials scientists. In assessing overall durability in a given environment not only must all the important effects be quantified individually but the individual effects need to be combined as well. This is a formidable task and it is only recently that these points have been considered seriously[1].

Over the past few years the durability of grc has been keenly debated within the grc industry and at conferences where the properties and new applications of the material are discussed in detail. We have seen in preceding chapters that some of the properties of grc made from the original Cem-FIL fibre, notably its strength and toughness, are significantly reduced with time in wet environments, and this deterioration takes place without any outward signs, at least at the outset[2,3]. A special symposium was held in the USA in 1985 to discuss the durability of grc[4]. Here we give our views.

Grc is a new family of materials and it will be a long time before the 'ultimate' properties of different types of grc are established from field experience. Since judgement is needed now about these ultimate properties in designing say, cladding panels or sewer liners, the only practical course of action is to develop some form of accelerated test. Such a test should be realistic from the viewpoint of the use of the material, and the results obtained in the test should be relatable to real time. The latter requires precise knowledge about the mechanisms that control the long-term properties of the material.

Over the years an acceptable form of accelerated test has been developed in the grc industry. Many years ago it was demonstrated at BRE that the rate of reduction in the strength and toughness of grc can be enhanced by increasing the temperature of water in which the material is kept. However, it was not until the development by Litherland and his colleagues[5] of the 'strand in cement' or SIC test that one could model glass/cement interactions under accelerated conditions realistically in terms of the properties of the fibre obtaining in those conditions. The details of the SIC test, the scientific principles on which it is based and its use in predicting the long-term bending strength of grc have been given in Chapters 1 and 4 and will not be repeated here. Here we examine the validity of the model that is used in these predictions. It is important to point out that only the bending strength of grc can be predicted reliably using the accelerated test procedures developed by Litherland *et al.*[5] Although empirical relationships can be constructed between flexural strength of grc in UK natural weather and the failure strain or impact behaviour of the composite[6], much further work is needed in this area. Until such time it will not be possible to comment meaningfully on the long-term viability of the grc material in all its manifold applications.

The predictive model of Litherland *et al.*[5] assumes that the reductions in the tensile and bending strength of grc over the long term in wet environments are entirely due to reductions in the strength of the fibre brought about by the corrosive chemical action of the cement. Over the years, various authors have pointed out that the rapid embrittlement of grc in wet conditions may, at least partially, be due to the filling up of glass fibre strands by the products of cement hydration. We have discussed this aspect of grc microstructure in detail in Chapter 8. More recently Bentur *et al.*[7] and Diamond[8] have gone further and suggested that the 'bundle filling' process is the main mechanism for the embrittlement as well as the strength reduction of wet aged grc at least during the first few years of its life. This view is currently receiving support among many scientists and engineers[4]. In view of the fact that a correct identification of the process(es) responsible for the deterioration of grc strength with time is essential from the point of view of durability prediction we discuss here the main arguments behind the two proposals.

Some examples have been given in Chapter 1 which show that, when reacted with an alkaline medium, uncoated AR glass fibres lose strength rapidly at temperatures higher than the ambient. The work of Proctor and Yale[9] has shown that commercially made glass fibre strands behave in the same way towards alkaline solutions. It has been known for a very long time[10] that zirconosilicate glasses, the source material for AR glass fibres, react chemically with alkaline solutions and the rate of this reaction is temperature dependent. We have seen in Chapter 1 that some mechanistic models have been proposed for such reactions. However, there is as yet no theory from which the strength of glass fibres subjected to these reactions can be predicted. The strength of a glass fibre depends very strongly on the physical state of its surface (i.e. nature and population of flaws), and changes in these features affected by chemical reactions are not easily measurable. In the work on single uncoated filaments of G20 composition (see Chapter 1) carried out at BRE there is evidence[11] that, if the fibres are handled with great care, reactions with alkaline solutions and cement extracts at ambient temperatures do not reduce the strength of AR glass fibres significantly at least up to one year. Some years ago Cohen and Diamond[12] had made a similar observation. However, if the uncoated filaments are not treated very carefully a sharp reduction in their strength occurs when they are kept in cement extract solutions over a period longer than a few months. The results obtained by Majumdar et al.[13] at 20°C and 65°C are reproduced in Figs 1.11(a) and 1.11(b). Commercially made AR glass fibre strands kept in highly alkaline solutions and in cement at ambient temperatures also show a reduction in strength[14,15]. It is not unreasonable to speculate that the corrosion of AR glass fibres by alkaline cements has a long incubation period.

An estimate of the effectiveness of fibre reinforcement in grc in any environment and after any age can be obtained by removing the fibre from the corresponding grc composites and measuring its strength. Some of the results obtained at BRE[16] for Cem-FIL fibres removed from different types of cement matrices at various times are shown in Fig. 9.1. It is clear that after ten years the fibres from OPC and HAC composites have retained significantly larger proportions of their initial strength in air storage than under wet conditions. This is in line with the trend in the long-term properties of these composites. The same is true of other results in Fig. 9.1, and generally speaking it is found that the composite strengths follow the trend of fibre strength in any particular environment. It can easily be argued that as non-Portland cements are less alkaline than OPC and as all cements are more alkaline when wet than dry because of the enhanced degree of hydration in the former condition, AR glass fibres retain the largest proportion of their initial strength when the environment is the least alkaline. Consequently, grc made from the least alkaline cement matrix retains the largest proportion of its initial strength when kept in a wet environment for a long time.

The results of Cohen and Diamond[12] on the strength of glass fibres extracted from OPC composites reported several years ago do not fully

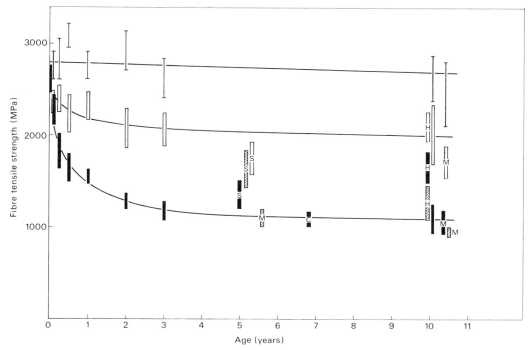

Fig. 9.1 Tensile strengths of individual AR glass fibres aged at normal temperatures. The curves are for G20 from OPC. The 'control' (strand chopped before OPC board manufacture and stored in a polythene bag) is represented by a vertical line ┃ ; OPC coupons stored in air ▯ , or water ▮ ; OPC mortar coupons containing 10% ground sand stored in air ▯$_M$, water ▮$_M$, or weathering ▨$_M$; supersulphate cement coupons stored in air ▯$_S$, water ▮$_S$, or weathering ▨$_S$; high alumina cement coupons stored in air ▯$_H$, water ▮$_H$, or weathering ▨$_H$; a coupon of 3 parts OPC:2 parts by weight of pfa cenospheres in water ▮$_C$.

The bars represent 95% confidence limits for the mean.

agree with the corresponding results shown in Fig. 9.1. In their experiments, AR glass fibres from a different source, when removed from OPC composites (kept at 22°C and 50% RH), did not show any significant reduction in strength up to about 500 days after the first 24 h of curing. The experimental technique used for removing fibres from grc composites at BRE was such that fibres from inside the fibre bundles were tested preferentially and not those on the outside and attached to the matrix as in the case of the study by Cohen and Diamond. There is thus some uncertainty as to whether the fibre strength values shown in Fig. 9.1 were typical. Whether this variation in technique can explain the differing trend in Fig. 9.1 from that reported by Cohen and Diamond is, however, subject to doubt. It is interesting to note that the initial strength of the AR glass fibres used by Cohen and Diamond lay in the range 1050–1380 MPa before mixing with cement, compared to more than twice this value shown in Fig. 9.1. Also, their strength value for the fibre after 500 days, 750 MPa, is considerably less than the lowest value shown in Fig. 9.1. It is also worth pointing out in Fig. 9.1 that fibre strengths for OPC composites reached nearly stable values only after three years in the wet-stored grc composites.

Proctor and Yale[9] have argued that glass fibre strand strength measured in the SIC test is likely to be the most realistic value of fibre strength in the grc composite. The SIC strength values of Cem-FIL AR glass fibres obtained by Litherland et al.[5] for various temperatures from 4° to 80°C are shown in Fig. 4.12. It is clear that at all temperatures fibre strand strength was reduced, and the rate of reduction increased sharply as the temperature increased. Since in these experiments the direct tensile strength of the fibre strand is measured, it is logical to assume that reduction in fibre strength is caused by glass/cement interaction. When different types of cements including non-Portland cements such as HAC or SSC were used in the SIC tests carried out at 50°C, Proctor and Yale[9] obtained the greatest reduction in fibre strength for the OPC matrix, which is considered to be the most alkaline among the cements used. It appears therefore that in these instances also there is a direct relationship between the strength of AR glass fibre strands in a cement matrix at any given time and temperature and the alkalinity of the matrix. In the SIC tests Cem-FIL 2 fibres retain a much higher proportion of their initial strength at any given time than the original Cem-FIL fibres[17,18] and it has been postulated that this is due to the much superior alkali resistance of the second generation AR glass fibres. The same comment can be made also for fibres such as AR Fibre-Super[19] provided with an inhibitive coating.

Experimental results on the long-term properties of grc given in preceding chapters clearly show that when the OPC matrix is blended with pozzolanas such as pfa and silica fume, or latently hydraulic materials such as ggbs, the long-term strength and toughness properties of the composite in wet conditions are sometimes superior to those of plain OPC/grc. In these cases the reactive materials added combine with Ca^{++} and alkali metal ions produced by the hydration of OPC, and one effect of this process is the reduction of OH^- concentrations in the pore solution (see Table 1.3). The other important effect is the production of the $C-S-H$ phase(s) in much greater amounts. This may reduce the permeability of the matrix significantly and thus limit the transport of OH^- and other harmful ions to the glass fibre surface. As long as particles of the reactive materials are present in the matrix to provide alternative reaction paths for the alkalis in cement, the glass fibre reinforcements will be exposed to a less hostile environment than would be the case otherwise. The fibre strengths are expected to be reduced to a much lesser extent over the long term in these blended cement matrices compared to OPC. These observations also suggest that there is a direct link between the long-term properties of grc and the alkalinity of the matrix.

Some of the observations described above can also be explained by assuming that precipitation of cement hydration products inside the glass fibre bundle is the key factor governing the reduction of some properties of grc with time. There is no doubt that in the case of grc made from non-Portland or blended cements, there is far less scope for large amounts of $Ca(OH)_2$ to precipitate in the inter-filamentary space of the glass fibre strand compared to the OPC composite. Some of these cements (HAC, for example) produce little if any $Ca(OH)_2$ on hydration, and consequently grc made from these cements would be expected to have better long-term properties than OPC/grc in wet environments, and indeed this is found to be the case. The results claimed for grc made from sulpho-aluminate cement[20] can also be explained in this way, and the same arguments can be extended to account for the much superior strength retention of grc in a relatively dry environment.

Another argument put forward in support of the 'bundle solidification' theory is the observation that grc modified by the inclusion of some polymer dispersions (e.g. acrylics) retains a much higher proportion of its strength and toughness in wet environments than is obtained with the unmodified composite. It is thought that the polymer particles occupy the inter-filamentary space in a glass fibre strand and thus prevent the precipitation of $Ca(OH)_2$ crystals inside the strand. Bijen[21] advanced such a notion to explain the long-term properties of acrylic polymer modified E-glass reinforced grc (Forton PGRC). Allen and Channer[22], on the other hand, had previously speculated that in these composites the polymer provided some protection to the E-glass fibre against the alkali attack of the cement matrices, but the results reported by Proctor et al.[17] do not support this view. Daniel and Schultz[18] have also shown in their polymer modified E-glass fibre reinforced cement specimens that the fibre surface was only partially covered by the polymer, and unprotected areas were etched by alkali attack.

If the solidification of the fibre bundle due to $Ca(OH)_2$ precipitation were the main governing factor for the reduction in the long-term strength and toughness of grc exposed to wet conditions, and if the polymer particles were able to prevent such a process by filling the inter-filamentary space in the strand, one would expect much better long-term durability of polymer modified grc in wet environments than has been observed so far in many cases[23,24]; see also Chapter 6. It is not clear to what extent the polymer film that may form around the cement grains limits the hydration of cement reducing the amount of Ca^{++} and OH^- ions, or in some specific cases how much of the Ca^{++} is removed from the solution by complexation with the polymer. It is very likely, as has been pointed out by Bijen[25] that the much reduced water absorption of polymer modified grc is the main reason for the better performance of the polymer modified materials compared to the unmodified composite.

According to some authors, Bijen[25] for example, the 'inhibitive coating' on Cem-FIL 2 prevents the crystallisation of $Ca(OH)_2$ inside the glass fibre bundles and gives the superior performance of Cem-FIL 2/grc over the composite made from the original Cem-FIL fibre in wet environments. It has been clear for some time[26] that some of the substances that were being patented as special constituents to be incorporated in the 'size' of the AR glass fibres in order to improve their performance as cement reinforcement were compounds that react easily with Ca^{++}. SEM photographs of fracture surfaces of Cem-FIL 2/grc composite (see Chapter 8) tend to indicate that even in long-term wet storage the fibre bundle in the composite remains integral and far fewer crystals of $Ca(OH)_2$ are present inside the fibre bundle than in the case of the corresponding samples of grc made from the original Cem-FIL fibre. It is important to point out that the polyvinyl acetate size on the original Cem-FIL fibre reacts easily with the Ca^{++} and the alkalis in cement, and the integrity of the strand is thus quickly lost. However, the difference in the microstructure of grc made from Cem-FIL and Cem-FIL 2 glass fibres is sufficient to be considered as an important factor in governing the relative durability of these composites. Proctor and his colleagues[27] have taken the view that the inhibitive coating on Cem-FIL 2 discourages the free hydration of cement near the fibre strand. They assert that the coating makes Cem-FIL 2 more alkali resistant than the original Cem-FIL fibre, otherwise its performance in the SIC tests, particularly at elevated temperatures, cannot be explained (see Fig. 4.15).

Several authors, for example Bentur[28], have used electron microscopy, particularly SEM, for the detection of flaws on the glass fibre surface that might have been formed by reaction with the cement alkali. From micrographs it is clear that, compared to the E-glass fibre surface, that of AR glass fibre extracted from grc appears to be very much cleaner, but it must be remembered that there are several types of flaws, some very small, that control the ultimate tensile strength of glass fibres[29]. The mere absence of visible surface flaws in SEM micrographs should not be taken as proof that the fibres have not reacted with the cement in a chemical sense and produced flaws. It must also be admitted that identification of features such as surface flaws is subject to specimen sampling and operator bias. Mills[30] has observed etch pits on AR glass fibres in two-year old grc samples.

The proponents of the bundle filling theory base their arguments for loss of strength and toughness in grc on the fracture mechanical notion of stress concentration at the crack tip, which in the case of the solidified bundle may be sufficiently strong to break the fibre after the matrix has cracked.

An increased strength of the bond between the fibre and the matrix in wet aged grc is considered to be a key factor in this argument. Aveston et al.[31] and others have pointed out that in brittle-matrix composites such as fibre reinforced cement, the fibres spanning a transverse crack at an angle are under considerable bending strain at points where they emerge from the crack face. Stucke and Majumdar[32] estimated that, in the case of grc reinforced by Cem-FIL fibres ($\sigma_f \sim 1800$ MPa) if the radius of curvature of the bent fibres is > 200 μm the fibres would not fracture, and they assumed that this condition prevails for the air-stored material with an open microstructure. In the old water-stored and naturally weathered material the much denser microstructure limits the radius of curvature to much smaller values, inducing the misaligned fibres to fail at the fracture face as illustrated in Fig. 9.2. The fracture initiation sites in the glass fibres can be seen in the micrograph which, as expected, are on the outer surfaces of the fibres at the position of maximum strain. The stress concentration on the fibre surface is significantly lessened if a compliant material is present in the neighbourhood of the fibre, and according to Bijen[25] such a mechanism is responsible for the better retention of strength by polymer modified grc compared to the unmodified material.

Even if it is assumed that stress concentration is an important factor in causing fracture of some fibres that are subjected to bending forces, such a phenomenon is unlikely to be of great significance when uniaxially reinforced grc composites are stressed in direct tension. As a result of precipitation in the fibre bundle, the total area of the fibre surface bonded to the matrix is

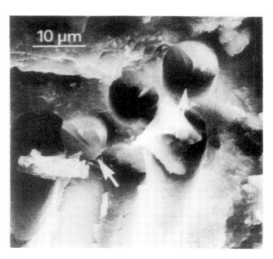

Fig. 9.2 Fracture initiation sites (indicated by arrows) in glass fibres bent at fracture surface in an upward direction.

no doubt greatly increased, but in these composites, according to the ACK theory, pull-out of fibres does not contribute to the ultimate tensile strength of the composite, which is given by $\sigma_f V_f$. It is doubtful if the interfacial bond strength can be greater than the shear strength of the matrix, which is low, and any stress concentration at the crack tip will be directed to debonding the fibre from the matrix rather than fracturing the fibre. Unfortunately there is very little information in the literature on the long-term tensile strength of continuously aligned grc composites kept in different environments. When such samples have been studied (see, for instance, Reference 7) MOR rather than UTS has been measured. The very limited unpublished results from various sources[14] available to us do show that the strength of grc undirectionally reinforced by the original Cem-FIL fibre is reduced with time in wet environments.

Fig. 9.3 shows the tensile stress—strain results obtained recently by Walton[15] on grc specimens in which about 3.4 vol% of Cem-FIL fibre strands were distributed throughout the thickness (7.6 mm) of the specimen as uniaxially, aligned, continuous reinforcement. It is seen that after 90 days in water at 19°C there is a significant reduction in the UTS and failure strain of the composite. In our view these losses in composite properties are

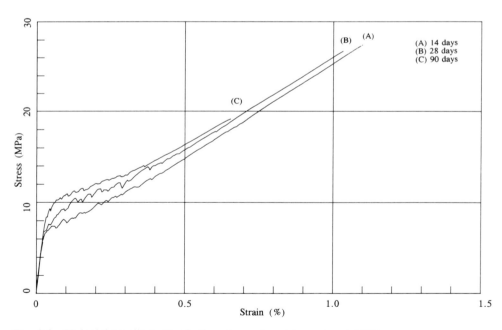

Fig. 9.3 Uniaxial tensile tests of aligned grc stored in water at 19°C.

mainly due to a reduction in the strength and failure strain of the fibre in the alkaline cement environment. Fig. 9.4 shows two SEM photographs of the microstructure of the 90 day specimen. The fibre bundle appears porous and lacking in cement hydration products. From these limited results it is obvious that for clarification of the mechanisms of strength loss in grc it is essential that more attention is given to aligned composites.

In wet-aged grc, which is a brittle material, the UTS is very close to the BOP value and the MOR to the corresponding LOP, and the fracture surfaces of these composites show a plain, non-fibrous appearance (Fig. 8.6). According to the fracture model for fibre cement composites proposed by Hannant et al.[33], in grc reinforced by Cem-FIL AR glass fibres the maximum stress in the fibre at matrix cracking is expected to be of the order of 290 MPa if a value of 3 MPa is assumed for the fibre/cement bond strength. Bond strength values of the order of 50 MPa will be required if the stress in the fibre is to exceed its breaking strength, say 1200 MPa. In view of the poor shear strength of the matrix it is doubtful whether such a strong bond can be sustained at the fibre/matrix interface in grc. The situation will change considerably if the tensile strength of the fibre in grc is assumed to decrease with time in wet environments.

In a brittle matrix composite, the amount of fibre must exceed a 'critical' volume in order to increase the strength and toughness of the material significantly. A reduction in the tensile strength of the fibre with time will produce a corresponding increase in the critical fibre volume. For spray-dewatered grc made by using short Cem-FIL glass fibres in an approximately random two-dimensional array, Hannant et al.[33] have calculated, from data in the extant literature on reduced fibre strength with time in cement, that a total fibre volume of more than 3.5% at 28 days and more than 7.4% after

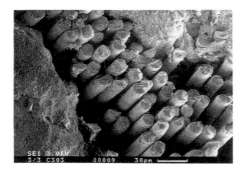 

Fig. 9.4 Microstructure of aligned grc.

ten years of weathering is required for a composite to maintain ductility. For aligned composites, the corresponding values are 1% at 28 days and 2% at ten years. It follows that, for a grc composite containing a fixed volume of fibres placed in a particular environment for a given length of time, the glass fibre strength has to exceed a certain 'critical' value for the fibre to function as an effective reinforcement, and this depends on the strength of the cement matrix, which increases with time. In a recent paper Proctor[34] has discussed in great detail the importance of these 'critical' fibre parameters in determining the stress–strain behaviour of grc with time in different environments. If the fibre volume fraction in grc is close to the 'critical' value at the time of manufacture, a modest reduction in fibre strength with time in any particular environment together with an increase in matrix cracking strength may produce a brittle behaviour in the composite which was once substantially pseudoductile, and the composites will fail by a single fracture. If fibre proportions in the composite are increased sufficiently, it is possible to delay, perhaps indefinitely, the total embrittlement of grc (see Fig. 4.9).

At BRE we have taken the view, mainly on microstructural grounds (Chapter 8), that precipitation of $Ca(OH)_2$ inside glass fibre bundles in grc is related, in some way or other, to the reduction in the strength, and particularly the impact resistance, of the composite in wet environments. Proctor and Yale[9] have shown that when placed under a slurry of $Ca(OH)_2$, Cem-FIL AR glass fibres become weaker with time relative to their strength in a saturated $Ca(OH)_2$ solution. If indeed $Ca(OH)_2$ crystals are instrumental in reducing the strength of the glass fibres in grc either by nucleation at pre-existing flaws on the fibre surface or by exerting crystallisation pressure or by some other mechanism, it may be possible to include these processes in the general corrosion model developed by Litherland et al.[5] In this respect the observation of Mills[35] that $Ca(OH)_2$ crystals are more easily precipitated on AR glass fibre surfaces in preference to other glasses, for example those derived from slag, may be important, although the role that the coating on the fibre might have played in these results remains unexamined. The tendency of $Ca(OH)_2$ to be attracted to zirconosilicate glasses in preference to silicate glasses of other compositions is difficult to explain on crystal chemical grounds.

It is generally understood that a part of the impact resistance of grc is due to fibre pull-out. In so far as precipitation within the fibre bundle will increase the difficulty of filaments slipping past each other, it is possible that some proportion of the reduction in the impact strength in wet environments is due to an increase in the fibre/matrix bond, and some of the ideas put forward by Bartos[36] may be relevant in this context. No model has been proposed yet that links the bundle filling mechanisms with the progressive reduction in the impact resistance of grc in wet environments, and long-term prediction is therefore not possible on this basis.

At the present time the chemical, alkali-corrosion model of Litherland et al.[5] is the only model available that can be used for predicting the bending and tensile strength of grc. In the final analysis the strength of the validity of this model is shown in Fig. 4.12, where it is seen that the strength of the fibre is reduced after reaction with cement, and in Fig. 9.5, which shows the linear relationship[37] between composite MOR and fibre strength and content ($\sigma_f v_f$) for a large number of grc specimens aged at 50°C and 35°C. The results shown in Fig. 9.5 add strong support to the chemical corrosion model.

The methodology developed by Litherland et al.[5] for predicting the long-term strength of grc from accelerated test results relies on an Arrhenius-type relationship to describe the temperature effect on grc strength. This methodology does not require any knowledge of the exact composition of the aqueous phase in the matrix or its alkalinity. Neither does it require any information about the properties of the matrix or those of the interface between the fibre and the matrix. It assumes, however, that the ultimate tensile strength of the composite is entirely dependent on fibre parameters

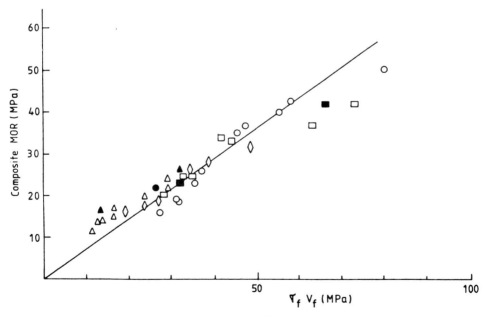

Fig. 9.5 Relation between composite MOR and fibre strength and content from accelerated ageing at 50°C and 35°C (Reference 37). Solid symbols from weathering at 1 and 10 years.

such as its strength, volume fraction, orientation, etc. − an assumption that is valid only for composites having fibre volumes in excess of the critical fibre volume for reinforcement. It has already been mentioned in a preceding chapter that according to the above model, Cem-FIL 2 and the original Cem-FIL glass fibres have been found to be equivalent.

Proctor et al.[17] have pointed out that the 'acceleration factors' derived by them for predicting the long-term properties of grc exposed to natural weather at various sites round the world (see Chapter 4) depend on the temperature dependence of the rate of strength loss in grc in these environments but not the rate of strength loss itself. The rate of strength loss, which can be very different in different types of grc, is controlled by factors that constitute the pre-exponential term of the Arrhenius relationship. It follows that provided the mechanism of strength reduction in grc is not altered by the addition of pozzolanas or other materials, it should still be possible to predict the long-term properties of these composites from accelerated test results using the Arrhenius-type relationship. In a complex system such as grc, the pre-exponential term may contain several parameters, and it is likely that concentration terms will be important in this respect. The principal effect of pozzolana addition, it may be argued, lies in the ability of these materials to control the concentration of hydroxyl ions at the glass fibre surface and thereby control the rate of fibre strength loss.

At temperatures higher than the ambient, say at 50°C, the rate of reaction between the pozzolana and cement will increase and it is likely that the strength of the matrix will also increase. But this will have little effect on the ultimate tensile strength of the composite provided that at the time of the observation the strength of the glass fibres has not diminished significantly and the fibres are still strong enough to carry the load when the matrix with improved strength has cracked.

In order to extend the validity of the use of the accelerating factors derived by Proctor et al. for grc to those made from heavily modified matrices it will be necessary to undertake further experimental work of the kind described by these authors using Cem-FIL or CEM-FIL 2 fibres as the reinforcement and the modified cements as matrices. It will also be necessary to obtain real-time strength results on these composites ranging at least up to ten years. For the time being predictions of the very long-term strength of composites of the types described above should be treated with caution.

Chapter Ten
Applications and Future Developments

Since 1972 AR-glass fibre reinforced cement (grc) has found its place as a versatile and commercially viable material for use in the construction industry. During this period it has also set the pace for the development of other fibre reinforced cement and concrete products. Starting with Biryukovich and co-workers[1] many scientists and engineers in different parts of the world have made important contributions in establishing and then extending the range of applications for which the material is suitable. In the early years of development of the grc industry in the UK some work was done at BRE by Ryder and others[2] to make prototypes of products such as cladding panels, pipes, window frames, fence posts, screens, etc. to demonstrate what could be done with the material. As the industry became more dynamic, the involvement of BRE in finding new applications became less prominent. However, in collaboration with Pilkington Brothers PLC and their licensees, BRE has kept a watching brief on new developments and applications of grc.

Many applications of grc have already been described by Young[3], True[4] and others. The proceedings of the Congresses organised by the Glassfibre Reinforced Cement Association include papers on many interesting products. Technical and illustrative literature issued by large firms such as Pilkington Brothers, Owens–Corning Fibreglass and Nippon Electric Glass also provides information on grc applications. In this chapter we give only a brief account of current uses and draw some conclusions about future developments.

Because of reduction in strength and toughness with time in wet environments, discussed in detail in earlier chapters, grc is not used in primary load-bearing or fully structural applications. The clearly prohibited applications are those where, if significant failures were to occur, there would be risks to life and limb or major loss and damage. BRE Digest 331[5] provides some guidelines and lists applications that are not recommended. This includes: (1) direct load-bearing frameworks, beams or columns; (2) major load-bearing wall units; (3) large or heavily profiled sandwich cladding panels; (4) suspended floor slabs; (5) major self-supporting roofs; (6) marine hulls and major self-supporting units.

10.1 Applications

10.1.1 Wall systems

Glass reinforced cement offers the architect and the engineer new and exciting opportunities in the design of wall systems. The sculptural nature of the material has encouraged innovation in shape, form and texture of the facades of buildings, and many examples now exist in various parts of the world. Grc cladding panels, both single-skin and sandwich, are of increasing relevance to industrial premises offering savings in transportation and erection costs.

Sandwich panels enable thermal insulation to be built into the wall elements, reducing heat loss and providing good fire resistance. Noise reduction and a more economic use of floor space can also be achieved by suitable panel design. Sandwich panels have been used in some countries, notably South Africa, to support the weight of lightweight roofing and ceiling elements in low-cost housing. Constructions of this kind are of great interest to developing nations.

One of the most important developments in the grc industry has been the adaptation of the 'stud frame' construction technique for building grc facades in high-rise office buildings[6]. Developed in the USA in the last decade, this particular form of construction is attracting wide interest at the present time. In this form of construction single skin grc cladding panels are attached to the stud frame by means of special joints.

The first use of grc on a large scale was in cladding systems, and some buildings in the UK clad with grc in the early 1970s have shown signs of distress (e.g. cracking and warping of the panel) requiring remedial actions. Moore[7] has discussed the problems that may be associated with the use of grc as a cladding material. In cases where the grc sheet on the exposed surface of sandwich panels showed cracking after several years of natural exposure, the material had behaved as predicted[8,9]. Moore pointed out, however, that sufficient attention had not been paid in design to the much reduced tensile strain at failure following weathering compared with that when the component was manufactured. It should perhaps be stated that the failures mentioned above all refer to grc using the original Cem-FIL glass fibres. It is expected that, by using the much improved Cem-FIL 2 AR glass fibres or its analogues and by including suitable polymers such as the acrylic dispersions in the matrix, the potential for cracking in grc cladding panels after long-term exposure will be significantly reduced or avoided. New cement formulations, for instance those being developed in West Germany, may offer further scope. Work is, however, still needed on the most appropriate core materials for sandwich panels.

10.1.2 Form work

Grc is particularly well suited to use as both permanent (lost) and temporary form work. The material has been used successfully in parapet shutters and bridge decking and in the construction of retaining walls, sewer linings and permanent coffer units. In view of the highly impermeable nature of grc, the Bridge Committee of the Norwegian Concrete Association has recommended that if 6 mm grc panels are used for bridge decking, the cover to the main steel reinforcement need only be 10 mm.

10.1.3 Pipework

The most notable example in this area is the slim line sewer pipe developed by Amey Roadstone Corporation Limited in the UK. The normal steel core of the spun concrete pipe has been replaced by inner and outer skins of grc produced *in situ* during manufacture of the pipe. Advantages of the slim line lie in the reduction in wall thickness and the use of an in-wall joint.

In some special applications grc has been used to provide protective layers for steel pipe lines.

10.1.4 Surface bonding and rendering

First pioneered by the US Department of Agriculture, grc surface bonding is a rapid method of wall construction in which a layer of cement mortar reinforced with suitable glass fibres is applied to the surfaces of a wall of dry masonry blocks that have been laid up dry. The base course of units needs to be laid in a full mortar bed, but mortar joints between subsequent courses can be eliminated. Either hand-trowelling or spraying can be used for the production of the grc layer, and similar techniques are applicable to one-coat surface renderings.

10.1.5 Agricultural uses

The combination of strength, light weight and ease of moulding to complex shapes has enabled grc to be used in a wide range of agricultural products. Examples are cattle drinking troughs, sheep dips, pig slurry channels, fish farming tanks, field drainage inspection chambers, etc. Grc has been used in the construction of industrial outbuildings used, for instance, in the intensive rearing of poultry.

10.1.6 Street, park and garden furniture

Advantage is taken in these applications of grc's light weight, non-combustibility and design flexibility in terms of shape, size and finish. Examples are street litter bins, flower planters and bench seats.

10.1.7 Refurbishment

Grc is being used increasingly in the repair and renovation of old concrete and stone buildings. The main advantages of the material in these applications are its mouldability, compatibility with existing materials such as concrete, its non-combustibility and ability to be formed in different colours and finishes. The ability of grc to simulate stone and brick has led to the production of a range of fireplace surrounds.

An interesting example of the use of grc in repair that is often cited[10] was carried out several years ago on freeze–thaw damaged concrete forming the walls of the Lower Monumental Lock and Dam on the Snake river in Washington State, USA. Ford[11] has described various aspects of this repair. A polymer modified grc formulation was used in the spray-up and the work was completed without any difficulty at a cost of only US $1.2 million compared with an estimate of US $26 million, the cost of conventional repair.

10.1.8 Asbestos-cement replacements

Because of the health hazards of asbestos[12] the use of the material in construction has decreased sharply in recent years in the UK and some other countries. In cement applications several different types of fibres (organic and inorganic), including alkali-resistant glass fibres, have been found suitable as replacements for asbestos. Nearly all major products of the asbestos-cement industry except pressure pipes have been made in grc on a commercial scale, and some of them, e.g. general purpose building boards and roofing slates, are on the market in the UK. Recently, corrugated grc sheet has been commercially produced in Germany and is being trial marketed. Glass fibre reinforced autoclaved calcium silicate is a good insulating and fire-resistant material.

10.1.9 Miscellaneous

The ease with which grc can be moulded together with the lightness of weight of many products and the attractive finishes that they can assume have made glass reinforced cement an extremely versatile material in the construction industry. These attributes have been exploited in many internal applications in public buildings, for example, as decorative ceiling panels for concert halls, wall lining of underground railway concourses and subways, and access flooring in modern office buildings.

Grc sheets can easily be formed into channels, and such components find application both inside and outside buildings. Grc sheets have been used in the construction of portable buildings such as bus shelters and kiosks and of holiday homes. Components for walkways and mooring pontoons in marinas have been built using grc, and vibratory driven grc piles have been used to protect canal banks. Noise barriers along motorways and railway lines have been built using grc. The non-combustible nature of the material has prompted its use in fire-check doors, sometimes with hollow pfa spheres known as 'cenospheres' included as a matrix component.

The range of vibrated and pressed products made from grc also indicates the versatility of the material. Vibration-cast screen wall panels are very popular in the Middle East. Pavement boxes and covers for gas stop valves are in general use. Conveyor belt covers and window frames, etc. have also been made from premixed grc. The grc promenade roof tile has been designed as a protective cover for built-up bituminous roofs to guard against damage to pedestrian traffic, to reduce solar heat gain and to provide safe access to chimneys and vented shafts.

10.2 Future prospects

In a relatively dry atmosphere, e.g. inside a building, grc is a very durable material and internal use will increase as new applications are identified. When used outdoors the high initial strength and toughness of the first generation grc has shown significant deterioration with time. The position has been substantially improved in recent years by the introduction of more alkali-resistant glass fibres, the development of inhibitive coatings and by the modification of the cement matrix. Experience with these second generation grc products will dictate how widely the material is going to be used in future. It may be necessary to develop new accelerated ageing tests for these new composites.

Probably the area of architectural cladding will remain the most controversial for the immediate future. In this context it should be remembered that alternative cladding materials such as concrete or glass fibre reinforced plastics have their own problems. The stud system of cladding using single-skin grc panels pioneered in the USA and now actively recommended for use in Australia, Japan, Singapore and other Far Eastern centres is expected to be developed in Europe on a bigger scale than hitherto. The future success of the grc industry may depend on the outcome of these trials.

For the forseeable future some concern will remain over the long-term performance of grc components outdoors. However, if the performance of the newer composites formed particularly from Cem-FIL 2 and polymer modified Portland cement prove as satisfactory as expected, a resurgence of interest in sandwich cladding panels might follow. There are several features of these panels such as their insulation properties and their suitability for wall and roof elements that are attractive for low cost housing. More care must, however, be taken in the design of the panels. The long-term properties as well as their cost-competitiveness will determine the future viability of the range of grc products already developed to replace asbestos cement products.

The use of special cements such as the sulphoaluminate or high alumina cements might increase if the applications justified the much higher cost of the cement compared to Portland cements. Sulphoaluminate cement has already made its mark in the grc industry in the Far East and there is now considerable interest in it in other countries. More work is necessary on matrix modifications by inorganic as well as organic additions, with a view to assisting in the manufacture of various grc products and/or improving their long-term performance in natural weather.

Research to improve the alkali resistance of glass fibre by changing its composition may continue but it is doubtful if great benefits can be secured in this way within economic limits unless a different and cheaper method of glass fibre production (e.g. by the sol-gel route) becomes a reality. There is much more scope in developing different types of 'inhibitive coating' for existing glass fibres.

Further work should be done to increase the understanding of the mechanism(s) for the loss in strength and durability of grc in wet environments. Until such understanding is secured and real-time long-term performance data are available for particular types of grc, it will be prudent not to use the material in primary structural elements. The restrictions given in *BRE Digest*[5] should be followed for the time being.

References

CHAPTER 1

1. **Nawy E. G., Neuwerth G. E. & Philips G. J.** (1970) *J. Struct Div.*, Am. Soc. Civ. Eng., **97**, 2203.
2. **Soames N. F.** (1963) Mag. Concr. Res., **15**, 151.
3. **Biryukovich K. L. & Biryukovich Yu. L.** (1961) *Stroit Mater.* (II) **1**, 18.
4. **Biryukovich K. L., Biryukovich Yu. L. & Biryukovich D. L.** (1965) *Glass Fibre Reinforced Cement*, Budivelnik, Kiev, 1964, Translation No. 12, Civ. Eng. Res Assoc., London.
5. **Romualdi J. P. & Batson G. B.** (1963) *J. Eng. Mech. Div.*, Proc. Am. Soc. Civ. Eng., **89**, 147.
6. **Krenchel H.** (1964) *Fibre Reinforcement*, Akademisk Forlag, Copenhagen.
7. **Dimbleby V. & Turner W. E. S.** (1926) *J. Soc. Glass Technol.*, **10**, 304.
8. **Majumdar A. J. & Ryder J. F.** (1986) *Glass Technol.*, **9**, 78.
9. **Majumdar A. J.** (1970) *Proc. Roy. Soc., London*, **A319**, 69.
10. **Urev N. B., Mikhailov N. V. & Rebinder P. A.** (1967) *Dokl. Acad. Nauk. SSSR* **177**, 1404.
11. **Lea F. M.** (1970) *The Chemistry of Cement and Concrete*, Edward Arnold (Publishers) Ltd.
12. **Robson T. D.** (1962) *High-Alumina Cements and Concretes*. Contractors Records Ltd.
13. **Longuet P., Burglen L. & Zelwer A.** (1973) *Rev. des Matér. de Constr. et de Trav. Publ.*, **676**, 35.
14. **Nixon P. J. & Page C. L.** (1986) Pore solution chemistry and alkali aggregate reaction. In *Proc. Katherine and Bryant Mather International Conference on Concrete Durability*, ACI SP-100, Vol. 2, p. 1833.
15. **Page C. L. & Vennesland O.** (1983) *Mater. and Constr.*, **16**, 19.
16. **Canham I., Page C. L. & Nixon P.J.** (1987) *Cem. and Concr. Res.*, **17**, 839.
17. **Majumdar A. J. & Nurse R. W.** (1974) *Mater. Sci. Eng.*, **15**, 107.
18. **Tanaka M. & Uchida I.** (1985) Application of the calcium silicates – $C_4A_3\bar{S}$ – slag type cement to GRC. In *Abstracts of 1985 Beijing Int. Symp. on Cement and Concrete*, China Academic Publishers, Beijing. p. 171.
19. **Shen Rongxi** (1986) Research and development of some new fibre reinforced cement composites in China. In *Proc. RILEM Symp. on Developments in Fibre Reinforced Cement and Concrete*. Sheffield, UK, July 1986, RILEM Tech. Committee 49 TFR, Vol. 2.
20. **Lowenstein K. L.** (1973) *The Manufacturing Technology of Continuous Glass Fibres*. Elsevier, Amsterdam.
21. **Proctor B. A. & Yale B.** (1980) *Phil. Trans. Roy. Soc., London*, **A 294**, 427.
22. **Shin W. S.** (1982) Recent development in alkali-resistant fibres at the Battelle Institute. In *Proc. Int. GRC Congress*, Paris, November 1981. Glassfibre Reinforced Cement Association, Newport, England, p. 359.
23. **Majumdar A. J.** (1985) Alkali-resistant glass fibres. In *Strong Fibres* (Ed. by W. Watt &

B. V. Perov), Chapter 2 in Handbook of Composites, Vol. 1, Elsevier Science Publishers BV.

24. **Fyles K., Litherland K. L. & Proctor B. A.** (1986) The effect of glass fibre compositions on the strength retention of grc. In *Proc. RILEM Symp. on Development in Fibre Reinforced Cement and Concrete*, Sheffield, UK, July 1986, RILEM Tech. Committee 49 TFR, Vol. 2.
25. **Gair H. G.** (1980) *Proc. Int. GRC Congress*, London, October 1979. Glassfibre Reinforced Cement Association, Newport, UK, p. 121.
26. UK Patent 1465059 (1977).
27. Health and Safety Commission (1979) *Asbestos*, Final Report of the Advisory Committee, Vol. 2, p. 20, HMSO London.
28. **Lawrence C. D.** (1966) Changes in composition of the aqueous phase during hydration of cement pastes and suspensions. In *Proc. Symp. on Structure of Portland Cement Paste and Concrete*, Special Report 90, Highway Research Board, Washington DC, p. 378.
29. **Litherland K. L., Maguire P. & Proctor B. A.** (1984) *Int. J. Cem. Composites and Lightweight Concr.*, **6**, 39.
30. **Proctor B. A.** (1986) The development and technology of AR fibres for cement reinforcement. In *Proc. Symp. on Durability of Glass Fibre Reinforced Concrete*, Chicago, November 1985, Prestressed Concrete Institute, p. 64.
31. **Larner L. J., Speakman K. & Majumdar A. J.** (1976) *J. Non-Cryst. Solids*, **20**, 43.
32. **Douglas R. W. & Isard J. O.** (1949) *J. Soc. Glass Technol.*, **33**, 289.
33. **Langford W. A., Davis K., Lamarche P., Laursen T., Groleau R. & Doremus R. H.** (1979). *J. Non-Cryst. Solids*, **33**, 249.
34. **Doremus R. H.** (1975) *J. Non-Cryst. Solids*, **19**, 137.
35. **Simhan R. G.** (1983) *J. Non-Cryst. Solids*, **54**, 335.
36. **Stucke M. S. & Majumdar A. J.** (1976) *J. Mater. Sci.*, **11**, 1019.
37. **Jaras A.** (Pilkington Brothers PLC) Personal communication.
38. **Chakroborty M., Das D., Basu S. & Paul A.** (1979) *Int. J. Cem. Composites*, **1**, 103.
39. **Li Chuanhe, Yicong Zou, Haolin Hou & Changman Yue** (1986) Research on titaniferrous alkali-resistant slag wool fibre cement. In *Proc. RILEM Symp. on Development in Fibre Reinforced Cement and Concrete*, Sheffield, UK, July 1986, RILEM Tech. Committee 49 TFR, Vol. 2.
40. **Vernon W. H. J., Akeroyd E. I. & Stroud E. J.** (1939) *J. Inst. Metals*, **65**, 301.
41. **Proctor B. A.** (1988) Pilkington Brothers PLC. Private communication.
42. **Majumdar A. J., West J. M. & Larner L. J.** (1977) *J. Mater. Sci.*, **12**, 927.
43. **Oakley D. R. & Proctor B. A.** (1975) Tensile stress-strain behaviour of glass fibre reinforced cement composites. In *Proc. RILEM Symp. on Fibre Reinforced Cement and Concrete*, London, September 1975, Construction Press, Lancaster p. 347.
44. **Cohen E. B. & Diamond S.** (1975) Validity of flexural strenth reduction as an indication of alkali attack on glass in fibre reinforced cement composites. In *Proc. RILEM Symp. on Fibre Reinforced Cement and Concrete*, London, September 1975, Construction Press, Lancaster, p. 315.

CHAPTER 2

1. **Krenchel H.** (1964) *Fibre Reinforcement*, Akademisk Forlag, Copenhagen.
2. **Allen H. G.** (1971) *J. Composite Mater.*, **5**, 194.
3. **Aveston J., Cooper G. A. & Kelly A.** (1971) Single and multiple fracture. In *Proc. NPL Conf. on The Properties of Fibre Composites*, November 1971, IPC Science and Technology Press, Guildford, UK, p. 15.
4. **Aveston J., Mercer R. A. & Sillwood J. M.** (1974) Fibre reinforced cements – scientific

foundations and specifications. In *Proc. NPL Conf. on Composites – Standards Testing and Design*, April 1974. IPC Science and Technology Press, Guildford, UK, p. 93.
5. **Proctor B. A.** (1990) A review of the theory of grc. *Cem. and Concr. Composites*, **12**, 53.
6. **Kimber A. C. & Keer J. G.** (1982) *J. Mater. Sci. Lett.*, **1**, 353.
7. **Laws V.** (1983) *J. Mater. Sci. Letts.*, **2**, 527.
8. **Hannant D. J.** (1975) The effect of post cracking ductility on the flexural strength of fibre cement and fibre concrete. In *Proc. RILEM Symp. on Fibre Reinforcement of Cement and Concrete*, London, September 1975, Construction Press, Lancaster, Vol. 2, p. 499.
9. **Bortz S. A.** (1969) *Structural Ceramic Composite Systems*, Chicago, IIT Research Institute AD692 149, p. 1.
10. **Laws V.** (1971) *J. Phys. D: Appl. Phys.*, **4**, 1737.
11. **de Vekey R. C. & Majumdar A. J.** (1968) *Mag. Concr. Res.*, **20**, 229.
12. **Laws V., Lawrence P. & Nurse R. W. B.** (1973) *J. Phys. D: Appl. Phys.*, **6**, 523.
13. **Laws V., Ali M. A. & Nurse R. W. B.** (1971) The response to stress of a short fibre reinforced brittle matrix. In *Proc. NPL Conf. on the Properties of Fibre Composites*, November 1971, IPC Science and Technology Press, Guildford, UK, p. 29.
14. **Hannant D. J.** (1978) *Fibre Cements and Fibre Concretes*, John Wiley and Sons, Chichester.
15. **Laws V.** (1987) *J. Mater. Sci.*, **6**, 675.
16. **Proctor B.** (1986) *J. Mater. Sci.*, **21**, 2441.
17. **Laws V. & Ali M. A.** (1977) The tensile stress/strain curve of brittle matrices reinforced with glass fibre. In *Proc. Conf. on Fibre Reinforced Materials, Design and Engineering Applications*. Institution of Civil Engineers, London, March 1977, p. 115.
18. **Laws V. & Walton P. L.** (1978) The tensile/bending relationship for fibre reinforced brittle matrices. In *Proc. RILEM Symp. on Testing and Test Methods of Fibre Cement Composites*, Sheffield, UK, April 1978, Construction Press, Lancaster, UK, p. 429.
19. **Hannant D. J., Hughes D. C. & Kelly A.** (1983) *Phil. Trans. Roy. Soc., London*, **A310**, 175.
20. **Majumdar A. J. & Laws V.** (1983) *Phil. Trans. Roy. Soc., London*, **A310**, 191.
21. **Romualdi J. P. & Batson G. B.** (1963) *J. Eng. Mech. Div.*, Proc. Am. Soc. Civ. Eng., **89**, 147.
22. **Kelly A.** (1970) *Proc. Roy. Soc., London*, **A 319**, 95.
23. **Greszcuk L. B.** (1969) Interfaces in Composites, ASTM STP452, p. 42.
24. **Lawrence P.** (1972) *J. Mater. Sci.*, **7**, 1.
25. **Bartos P.** (1980) *J. Mater. Sci.*, **15**, 3122.
26. **Laws V.** (1982) *Composites*, **13**, 145.

CHAPTER 3

1. **Blake H. V.** (1982) The competitiveness of grc with other materials. In *Proc. Int. GRC Congr.*, Paris, November 1981. Glass Fibre Reinforced Cement Association, Newport, UK, p. 327.
2. **True G.** (1986) *GRC Production and Uses*, Palladian Publications, London.
3. **Young J.** (1978) *Designing with GRC*, Architectural Press, London.
4. **Smith J. W.** (1986) *A Review of GRC Production Processes*. Glass Fibre Reinforced Cement Association, Newport, UK.
5. **Pilkington** (1985) *Cem-FIL GRC Technical Data*. Pilkington Reinforcements Limited, St Helens, UK.
6. **Ryder J. F.** (1975) Applications of fibre cement. In *Proc. RILEM Symp. on Fibre Reinforced Cement and Concrete*, London, September 1975. Construction Press,

Lancaster, UK, p. 23.
7. **Hills D. L.** (1975) *Precast Concr.*, **6**, 251.
8. **Edens G.** (1988) GRC profile spray machine for continuous or semi-continuous section production. In *Proc. Int. GRC Congr.*, Edinburgh, October 1987. Glassfibre Reinforced Cement Association, Newport, UK, p. 115.
9. **Mishima K., Kozuka M., Ishizuka T. & Takeda R.** (1980) High throughput grc sheet production technology. In *Proc. Int. GRC Congr.*, London, October 1979. Glass Fibre Reinforced Cement Association, Newport, UK, p. 93.
10. Concrete Society (1973) Technical Report No. 51.067. British Cement Association, Slough, UK.
11. **Smith J. W.** (1982) *Composites*, **13**, 161.
12. **Ball H. P.** (1984) The effect of Forton compound on gfrc curing requirements. In *Proc. Int. GRC Congr.*, Stratford-on-Avon, UK, October 1983. Glass Fibre Reinforced Cement Association, Newport, UK, p. 56.
13. **Ward D. & Proctor B. A.** (1978) Quality control test methods for glass fibre reinforced cement. In *Proc. RILEM Symp. on Testing and Test Methods of Fibre Cement Composites*, Sheffield, UK, April 1978. Construction Press, Lancaster, UK, p. 35.
14. **Hibbert A. P.** (1974) *J. Mater. Sci.*, **9**, 512.
15. **Hibbert A. P. & Grimer F. J.** (1975) *J. Mater. Sci.*, **10**, 2124.
16. **Rayment D. L. & Majumdar A. J.** (1978) *J. Mater. Sci.*, **13**, 817.
17. **Ashley D. G.** (1978) Measurement of glass content in fibre cement composites by x-ray fluorescence analysis. In *Proc. RILEM Symp. of Testing and Test Methods of Fibre Cement Composites*, Sheffield, UK, April 1978. Construction Press, Lancaster, UK, p. 265.

CHAPTER 4

1. Cem-FIL GRC Technical Data (1985) Pilkington Brothers PLC, St Helens, Merseyside.
2. **Ali M. A., Majumdar A. J. & Singh B.** (1975) *J. Mater. Sci.*, **10**, 1732.
3. **Majumdar A. J., Singh B., Langley A. A. & Ali M. A.** (1980) *J. Mater. Sci.*, **15**, 1085.
4. **Singh B. & Majumdar A. J.** (1985) *J. Mater. Sci. Lett.*, **4**, 967.
5. **Singh B. & Majumdar A. J.** (1987) *Int. J. Cem. Comp. and Lightwt. Concr.*, **9**, 75.
6. Building Research Establishment (1979) *Properties of GRC: Ten Year Results*. BRE Information Paper IP 36/79.
7. Building Research Establishment (1976) A Study of the Properties of *Cem-FIL/OPC Composites*. BRE Current Paper CP 38/76.
8. **Ohta H.** (1986) New AR glassfibre 'AR fibre super'. In *Proc. Int. GRC Congress*, Darmstadt, W. Germany, October 1985. Glassfibre Reinforced Cement Association, Newport, UK, p. 37.
9. **Ferry R.** (1988) The effect of the strand length and glass fibre content on the properties of Cem-FIL 2 grc. In *Proc. Int. GRC Congr.*, Edinburgh, October 1987. Glass Fibre Reinforced Cement Association, Newport, UK, p. 101.
10. **Litherland K. L., Oakley D. R. & Proctor B. A.** (1981) *Cem. & Concr. Res.*, **11**, 455.
11. **Proctor B. A., Oakley D. R. & Litherland K. L.** (1982) *Composites*, **13**, 173.
12. **Litherland K. L.** (1986) Test methods for evaluating the long-term behaviour of gfrc. In *Proc. Symp. on Durability of Glass Fibre Reinforced Concrete*, Chicago, November 1985. Prestressed Concrete Institute, Chicago, p. 210.
13. **Hills D. L.** (1975) *Precast Concr.*, **6**, 251.
14. **Langley A. A.** (1981) *Mag. Concr. Res.*, **33**, 221.
15. **Allen H. G. & Jolly C. K.** (1982) *J. Mater. Sci.*, **17**, 2037.
16. **Proctor B. A.** (1980) Properties and performance of grc. In *Fibrous Concrete*, Proc. Concr. Soc. Symp. on Fibrous Concrete, London, April 1980. Construction Press,

Lancaster, UK, p. 69.
17. **Hibbert A. P. & Grimer F. J.** (1975) *J. Mater. Sci.*, **10**, 2124.
18. **West J. M. & Walton P. L.** (1981) *J. Mater. Sci.*, **16**, 2398.
19. **Warrior D. A. & Rothwell K. P.** (1982) Chemical resistance of grc. In *Proc. Int. GRC Congress*, Paris, November 1981. Glassfibre Reinforced Cement Association, Newport, UK, p. 100.

CHAPTER 5

1. **Singh B. & Majumdar A. J.** (1981) *Int. J. Cem. Composites and Lightwt. Concr.*, **3**, 93.
2. **Singh B., Majumdar A. J. & Ali M. A.** (1984) *Int. J. Cem. Composites and Lightwt. Concr.*, **6**, 65.
3. **Singh B. & Majumdar A. J.** (1985) *Int. J. Cem. Composites and Lightwt. Concr.*, **7**, 3.
4. **Proctor B. A., Oakley D. R. & Litherland K. L.** (1982) *Composites*, **13**, 173.
5. **Majumdar A. J., Evans T. J. & Singh B.** (1986) The properties of Cem-FIL 2/grc. In *Proc. Int. GRC Congr.*, Darmstadt, W. Germany, October 1985. Glassfibre Reinforced Cement Associaton, Newport, UK, p. 79.
6. **Anon. Northwich Building Board** (1984) *Asbestos Substitute Cuts Out Health Hazard. Surveyor*, 23 Feb., 12.
7. **Hayashi M., Sato S. & Fuji H.** (1986) Some ways to improve durability of GFRC. In *Proc. Symp. on Durability of Glass Fibre Reinforced Concrete*, Chicago, November 1985. Precast Concrete Institute, Chicago, p. 270.
8. **Bentur A. & Diamond S.** (1986) Effects of direct incorporation of microsilica into gfrc composites on retention of mechanical properties after ageing. In *Proc. Symp. on Durability of Glass Fibre Reinforced Concrete*, Chicago, November 1985. Precast Concrete Institute, Chicago, p. 337.
9. **Ambroise J., Dejean J., Foumi J. & Pera J.** (1988) Metakaolin blended cement – an efficient way to improve grc durability. In *Proc. Int. GRC Congr.*, Edinburgh, October 1987. Glassfibre Reinforced Cement Association, Newport, UK, p. 19.
10. **West J. M., Majumdar A. J. & de Vekey R. C.** (1980) *Composites*, **11**, 19.
11. Building Research Establishment (1981) *Fibre Reinforced Lightweight Inorganic Materials*. BRE Information Paper IP 29/81.
12. **West J. M., Majumdar A. J. & de Vekey R. C.** (1980) *Composites*, **11**, 169.

CHAPTER 6

1. **Synthetic resins in building construction**. In *Proc. RILEM Symp. 1967*, Eyrolles Paris (1970).
2. **Steinberg M., Kukacka M., Colombo L. E.** *et al.* (1968) *Concrete-Polymer Materials*. First Topical Report BNL 50134 (T-509), Brookhaven National Laboratory.
3. Concrete Society Technical Report No. 9. *Polymer Concretes*. Report of a working party, Concrete Society, London (1975).
4. **Biryukovich K. L., Biryukovich Yu. L. & Biryukovich D. L.** (1965) *Glass Fibre Reinforced Cement*, Budivelnik, Kiev, 1964, Translation No. 12, Civ. Eng. Res. Assoc., London.
5. **de Vekey R. C. & Majumdar A. J.** (1975) *Mater. Struct.* **8**, 315.
6. **de Vekey, R. C.** (1976) The properties of polymer modified cement pastes. In *Proc. Int. Conf. on Polymer Concrete*, London, May 1975. Construction Press, Lancaster, UK, p. 97.

7. **Majumdar A. J.** (1974) Modification of grc properties. In *Proc. NPL Conf. on Composites – Standards Testing and Design*, April 1974. IPC Science and Technology Press, Guildford, UK, p. 108.
8. **West J. M., de Vekey R. C. & Majumdar A. J.** (1985) *Composites*, **16**, 33.
9. **West J. M., Majumdar A. J. & de Vekey R. C.** (1986) *Composites*, **17**, 56.
10. **Majumdar A. J., Singh B. & West J. M.** (1987) *Composites*, **18**, 61.
11. Building Research Establishment (1979) *Properties of GRC: 10 Year Results*, BRE Information Paper IP 36/79.
12. **Green M. F., Oakley D. R. & Proctor B. A.** (1978) Tensile testing of glass reinforced cement sheet. In *Proc. RILEM Symp. on Testing and Test Methods of Fibre Cement Composites*, Sheffield, UK, April 1978. Construction Press, Lancaster, UK, p. 439.
13. **Ball H. P.** (1984) The effect of Forton compound on gfrc curing requirements. In *Proc. Int. GRC Congr.*, Stratford–on–Avon, October 1983. Glass Fibre Reinforced Cement Association, Newport, UK, p. 56.
14. **Knowles, R. P. & Proctor B. A.** (1988) The properties and performance of polymer modified grc. In *Proc. Int. GRC Congr.*, Edinburgh, October 1987. Glass Fibre Reinforced Cement Association, Newport, UK, p. 79.
15. **Bijen J.** (1988). Curing of grc. In *Proc. Int. GRC Congr.*, Edinburgh, October 1987. Glass Fibre Reinforced Cement Association, Newport, UK, p. 71.
16. **Daniel J. I. & Pecoraro M. E.** (1982) *Effect of Forton Polymer on Curing Requirements of AR Glass Fibre Reinforced Cement Composites*. Portland Cement Association Construction Technology Laboratories, October 1982.
17. **Allen H. G. & Channer R. S.** (1976) Some mechanical properties of polymer modified Portland cement sheets with and without glass fibre reinforcement. In *Proc. Int. Conf. on Polymer Concrete*, London, May 1975. Construction Press, Lancaster, UK, p. 282.
18. **Bijen J.** (1980) E-glass fibre reinforced polymer modified cement. In *Proc. Int. GRC Congr.*, London, October 1979. Glass Fibre Reinforced Cement Association, Newport, UK, p. 62.
19. **Jacobs M. J. N.** (1986) Durability of pgrc, design implications. In *Proc. Int. GRC Congr.*, Darmstadt, W. Germany, October 1985. Glass Fibre Reinforced Cement Association, Newport, UK, p. 53.
20. **Bijen J.** (1986) A survey of new developments in glass composition, coatings and matrices to extend service lifetime of gfrc. In *Proc. Symp. on Durability of Glass Fibre Reinforced Concrete*. Chicago, November 1985. Prestressed Concrete Institute, Chicago, p. 251.
21. **Proctor B. A., Oakley D. R. & Litherland K. L.** (1982) *Composites*, **13**, 173.
22. **Daniel J. I. & Schultz D. M.** (1986) Durability of glass fibre reinforced concrete systems. In *Proc. Symp. on Durability of Glass Fibre Reinforced Concrete*. Chicago, November 1985. Prestressed Concrete Institute, Chicago, p. 174.
23. **Allen H. G.** (1975) *Glass Fibre Reinforced Cement – Strength and Stiffness*. Civ. Eng. Res. and Inf. Assoc. Report No. 55, CIRIA, London.
24. **Grimer F. G. & Ali M. A.** (1969) *Mag. Concr. Res.*, **21**, 23.
25. **Briggs A. & Ayres C. F.** (1975) The strength of polymer impregnated glass fibre reinforced Portland cement. In *Proc. RILEM Symp. on Fibre Cement and Fibre Concrete*. London, September 1975. Construction Press, Lancaster UK, Vol. 2, p. 601.

CHAPTER 7

1. **Biryukovich K. L., Biryukovich Yu. L. & Biryukovich D. L.** (1965) *Glass Fibre Reinforced Cement*, Budivelnik, Kiev, 1964. Translation No. 12, Civ. Eng. Res. Assoc., London.
2. **Majumdar, A. J., Singh B. & Ali M. A.** (1981) *J. Mater. Sci.*, **16**, 2597.
3. **Majumdar A. J., Singh B. & Evans T. J.** (1981) *Composites*, **12**, 177.

4. **Singh B. & Majumdar A. J.** (1987) *Composites*, **18**, 329.
5. **Singh B., Walton P. L. & Stucke M. S.** (1978) Test methods used to measure the mechanical properties of fibre-cement composites at BRE. In *Proc. RILEM Symp. on Testing and Test Methods of Fibre Cement Composites*, Sheffield, UK, April 1978, Construction Press, Lancaster, UK, p. 377.
6. **Bate S. S.** (1974) *Report on the Failures of Roof Beams at Sir John Cass' Foundation and Red Coat Church of England Secondary School, Stepney*, BRE Current Paper CP58/74.
7. **Neville A. M.** (1975) *High Alumina Cement Concrete*. Construction Press, Lancaster, UK.
8. Building Research Establishment (1979) *Properties of GRC: Ten Year Results*. BRE Information Paper, IP36/79.
9. **Proctor B. A. & Yale B.** (1980) *Phil. Trans. Roy. Soc., London*, **A294**, 427.
10. **Midgley H. G.** (1967) *Trans. Brit. Ceram. Soc.*, **66**, 161.
11. **Midgley H. G. & Pettifer K.** (1972) *Trans. Brit. Ceram. Soc.*, **71**, 55.
12. **Proctor B. A. & Litherland K.** (1986) Improving the strength retention of grc by matrix and fibre modifications. In *Proc. Int. GRC Congr.*, Darmstadt, W. Germany, October 1985. Glassfibre Reinforced Cement Association, Newport, UK, p. 45.
13. **Majumdar A. J. & Stucke M. S.** (1981) *Cem. Concr. Res.*, **11**, 781.
14. UK Patent No. 1214779 (1970).
15. **Shen Rongxi** (1986) Research and development of some new fibre reinforced cement composites in China. In *Proc. RILEM Symp. on Developments in Fibre Reinforced Cement and Concrete*. Sheffield, UK, July 1986,. RILEM Tech. Committee 49, TFR Vol. 2.
16. **Hyashi M., Sato S. & Funjii H.** (1986) Some ways to improve gfrc. In *Proc. Symp. on Durability of Glass Fibre Reinforced Concrete*, Chicago, November 1985. Prestressed Concrete Institute, Chicago, p. 270.
17. **West J. M., Speakman K. & Majumdar A. J.** (1974) *Glass Fibre Reinforced Autoclaved Calcium Silicate Insulation Material*, BRE Current Paper CP 62/74.
18. **Majumdar A. J. & West J. M.** (1981) *Fibre Reinforced Lightweight Inorganic Materials*, BRE Information Paper IP 29/81.
19. **Clarke L. L.** (1975) *Reinforcement of Cementitious Materials with Glass Fibres*. Owens Corning Fibreglass, Granville, Ohio.
20. **Speakman K. & Majumdar A. J.** (1973) *Health Aspects of an Asbestos Board Substitute*. BSRA member firms' conference relating to accomodation bulkheads, June 1973. British Ship Research Association, Wallsend Research Station, Wallsend, England.

CHAPTER 8

1. **Jaras A. C. & Litherland K. L.** (1975) Microstructural features of glass fibre reinforced cement composites. *Proc. RILEM Symp. on Fibre Reinforced Cement and Concrete*, London, September 1975. Construction Press, Lancaster, UK, p. 327.
2. **Bentur A. & Diamond S.** (1984) *Cem. and Concr. Res.*, **14**, 31.
3. **Stucke M. S. & Majumdar A. J.** (1976) *J. Mater. Sci.*, **11**, 1019.
4. **Laws V., Langley A. A. & West J. M.** (1986) *J. Mater. Sci.*, **21**, 289.
5. **Bentur A., Ben Bassat M. & Schneider O.** (1985) *J. Am. Ceram. Soc.*, **68**, 203.
6. **Pinchin D. and Tabor D.** (1978) *Cem. and Concr. Res.*, **8**, 15.
7. **Barnes B. D., Diamond S. & Dolch W. L.** (1978) *Cem. and Concr. Res.*, **8**, 233.
8. **Mills R. H.** (1981) Preferential precipitation of calcium hydroxide on alkali-resistant glass fibres. In *Proc. Mater. Res. Soc. Symp. on Advances in Cement Matrix Composites*, Boston, November 1980, p. 229.
9. **Fyles K., Litherland K. L. & Proctor B. A.** (1986) The effect of glass fibre compositions on the strength retention of grc. In *Proc. RILEM Symp. on Development in Fibre*

Reinforced Cement and Concrete, Sheffield, UK, July 1986, RILEM Tech, Committee 49, TFR Vol. 2.
10. **Majumdar A. J. & Stucke M. S.** (1981) Cem. and Concr. Res., **11**, 781.
11. **Aveston J., Cooper G. A. & Kelly A.** (1971) Single and multiple fracture. In Proc. NPL Conf. on the Properties of Fibre Composites, November 1971, IPC Science and Technology Press, Guildford, p. 15.
12. **Oakley D. R. & Proctor B. A.** (1975) Tensile stress-strain behaviour of glass fibre reinforced cement composites. In Proc. RILEM Symp. on Fibre Reinforced Cement and Concrete, London, September 1975. Construction Press, Lancaster UK, p. 347.
13. **de Vekey R. C. & Majumdar A. J.** (1968) Mag. Concr. Res., **20**, 229.
14. **de Vekey R. C. & Majumdar A. J.** (1970) J. Mater. Sci. Lett., **5**, 183.
15. **Bartos P.** (1981) Int. J. Cem. Composites and Lightweight Concr., **3**, 159.

CHAPTER 9

1. **Masters L. W. & Brandt E.** (1987) Prediction of service life of building materials and components. Final report CIB W80/RILEM 71 PSL, Mater. Struct., **20**, 55.
2. **Moore J. F. A.** (1984) The use of glass reinforced cement in cladding panels. BRE Report.
3. Building Research Establishment Digest 331, GRC (1988).
4. **Prestressed Concrete Institute** (1986) Proc. Symp. on Durability of Glass Fibre Reinforced Concrete, Chicago, November 1985, PCI.
5. **Litherland K. L., Oakley D. R. & Proctor B. A.** (1981) Cem. and Concr. Res., **11**, 455.
6. **Litherland K. L.** (1986) Test methods of evaluating the long-term behaviour of grc. Proc. Symp. on Durability of Glass Fibre Reinforced Concrete, Chicago, November 1985, PCI, p. 210.
7. **Bentur A., Ben Bassat M. & Schnider O.** (1985) J. Am. Ceram. Soc., **68**, 203.
8. **Diamond S.** (1986) A summary and restrospective of the symposium on durability of gfrc. Proc. Symp. on Durability of Glass Fibre Reinforced Concrete, Chicago, November 1985, PCI, p. 352.
9. **Proctor B. A. & Yale B.** (1980) Phil. Trans. Roy. Soc., London, **A294**, 427.
10. **Dimbleby V. & Turner W. E. S.** (1926) J. Soc. Glass Technol., **10**, 304.
11. **West J. M.** (1977) Unpublished report, Building Research Establishment.
12. **Cohen E. B. & Diamond S.** (1975) Validity of flexural strength reduction as an indication of alkali attack on glass in fibre reinforced cement composites. In Proc. RILEM Symp. on Fibre Reinforced Cement and Concrete, London, September 1975. Construction Press, Lancaster, UK, p. 315.
13. **Majumdar A. J., West J. M. & Larner L. J.** (1977) J. Mater. Sci., **12**, 927.
14. **Proctor B. A.** (1988) Private communication.
15. **Walton P. L.** (1990) Private communication.
16. **West J. M. & Majumdar A. J.** (1982) J. Mater. Sci. Lett., **1**, 214.
17. **Proctor B. A., Oakley D. R. & Litherland K. L.** (1982) Composites, **13**, 173.
18. **Daniel J. I. & Schultz D. M.** (1986) Durability of glass fibre reinforced concrete systems. Proc. Symp. on Durability of Glass Fibre Reinforced Concrete, Chicago, November 1985, PCI, p. 174.
19. **Ohta H., Yanagisawa O., Mukaiyama T., Ohigashi T. & Takeda R.** (1986) New AR glass fibre, 'AR Fibre-Super'. In Proc. Int. GRC Congr., Darmstadt, W. Germany, October 1985. Glass Fibre Reinforced Cement Association, Newport, UK, p. 37.
20. **Tanaka M. & Uchida I.** (1986) Durability of GFRC with calcium silicate $- C_4A_3\bar{S} - C\bar{S} -$ slag type low alkaline cement. Proc. Symp. on Durability of Glass Fibre Reinforced

Concrete, Chicago, November 1985, PCI, p. 304.
21. **Bijen J.** (1980) E-glass fibre reinforced polymer modified cement. In *Proc. Int. Congr. on Glass Fibre Reinforced Cement*, London, October 1979. Glass Fibre Reinforced Cement Association, Newport, UK, p. 62.
22. **Allen H. G. & Channer R. S.** (1976) Some mechanical properties of polymer modified Portland cement sheets with or without glass fibre reinforcement. In *Proc. Int. Congr. on Polymer Concrete*, London, May 1975. Construction Press, Lancaster, UK, p. 282.
23. **Majumdar A. J. & West J. M.** (1987) *Polymer modified grc*. BRE Information Paper IP 10/87.
24. **Cimilli T.** (1986) Durability of glass fibres in polymer modified cement. In *Proc. Symp. on Durability of Glass Fibre Reinforced Concrete*, Chicago, November 1985, PCI, p. 315.
25. **Bijen J.** (1986) A survey of new developments in glass composition, coatings and matrices to extend service lifetime of gfrc. In *Proc. Symp. on Durability of Glass Fibre Reinforced Concrete*, Chicago, November 1985, PCI, p. 251.
26. **Majumdar A. J.** (1981) Some aspects of glass fibre reinforced cement research. *Proc. Mater. Res. Soc. Symp. L on Advances in Cement-Matrix Composites*, Boston, November 1980, p. 37.
27. **Proctor B. A.** (1986) The development and technology of AR fibres for cement reinforcement. In *Proc. Symp. on Durability of Glass Fibre Reinforced Concrete*, Chicago, November 1985, PCI, p. 64.
28. **Bentur A.** (1986) Mechanisms of potential embrittlement and strength loss of glass fibre reinforced cement composites. In *Proc. Symp. on Durability of Glass Fibre Reinforced Concrete*, Chicago, November 1985, PCI, p. 109.
29. **Schmitz G. K.** (1965) *Exploration and Evaluation of New Glasses in Fibre Form*. Clearinghouse AD, 464 261.
30. **Mills R. H.** (1981) *Cem. and Concr. Res.*, **11**, 421.
31. **Aveston J., Mercer R. A. & Sillwood J. M.** (1974) Fibre reinforced cements – scientific foundations and specifications. In *Proc. NPL Conf. on Composites – Standards Testing and Design*, April 1974, IPC Science and Technology Press, Guildford, UK, p. 93.
32. **Stucke M. S. & Majumdar A. J.** (1976) *J. Mater. Sci.*, **11**, 1019.
33. **Hannant D. J., Hughes D. C. & Kelly A.** (1983) *Phil. Trans. Roy. Soc. London*, **A310**, 175.
34. **Proctor B. A.** (1986) *J. Mater. Sci.*, **21**, 2441.
35. **Mills R. H.** (1981) *Cem. and Concr. Res.*, **11**, 689.
36. **Bartos P.** (1986) Effects of changes in fibre strength and bond characteristics due to ageing on fracture mechanisms of grc. In *Proc. Symp. on Durability of Glass Fibre Reinforced Concrete*, Chicago, November 1985, PCI, p. 136.
37. **Litherland K. L. & Proctor B. A.** (1986) The effect of matrix formulation, fibre content and fibre composition on the durability of glass fibre reinforced cement. In *Proc. Symp. on Durability of Glass Fibre Reinforced Concrete*, Chicago, November 1985, PCI, p. 124.

CHAPTER 10

1. **Biryukovich K. L., Biryukovich Yu L. & Biryukovich D. L.** (1965) *Glass Fibre Reinforced Cement*. Budivelinik, Kiev, 1964; Translation No. 12, Civil Eng. Res. Assoc., London.
2. **Ryder J. F.** (1975) Applications of fibre cement. In *Proc. RILEM Symposium on Fibre Reinforced Cement and Concrete*, London, September 1975. Construction Press, Lancaster, UK, p. 23.
3. **Young J.** (1978) *Designing with GRC*. Architectural Press, London.
4. **True G.** (1986) *GRC Production and Uses*. Palladian Publications Ltd, London.

5. Building Research Establishment Digest 331, GRC (1988).
6. **Molloy B.** (1988) GFRC stud-frame cladding in the USA. *Proc. Int. GRC Congr.*, Edinburgh, October 1987, Glassfibre Reinforced Cement Association, Newport, UK, p. 263.
7. **Moore J. F. A.** (1984) The Use of Glass-Reinforced Cement in Cladding Panels. BRE Information Paper, IP 5/84.
8. Building Research Establishment (1976) *A Study of the Properties of Cem-FIL/OPC Composites*. BRE Current Paper, CP 38/76.
9. Building Research Establishment (1979) *Properties of GRC: 10 year results*, BRE Information Paper, IP 36/79.
10. **Dunstan I.** (1984) Keynote Address. In *Proc. Int. GRC Congr.*, Stratford-upon-Avon, October 1983. Glassfibre Reinforced Cement Association, Newport, UK, p. 3.
11. **Ford J. B.** Repair of freeze-thaw damaged concrete with grc latex spray-up. In *Proc. Int. GRC Congr.*, Paris, November 1981, Glassfibre Reinforced Cement Association, Newport, UK, p. 176.
12. Health and Safety Commission (1979) *Asbestos*. Final Report of the Advisory Committee, Vols. 1 and 2. HMSO London.

Index

accelerated
  ageing, 82, 102, 105, 111, 128
  tests, 82, 84–6, 98, 100, 107, 111, 125, 139, 165, 175
acceleration factor, 176
  see also equivalent factor
accelerators, 56, 66
acid resistance, 91, 134
ACK model, 27, 28, 49, 172
acrylic(s), 6, 66, 112, 123–5, 169
activation energy, 19, 84
admixture, 2, 55, 62
air entraining agent, 56
alkali attack, 20, 165, 169
alkali-resistance, tests, 15–17
alkalinity, 7, 9, 92, 168
analysis, dispersive X-ray (EDXA), 20
anisotropy, 100
applications, 177–81
AR slag wool, 13, 21
Arrhenius relationship, 19, 82, 175, 176
Asahi Glass, 59, 82
asbestos, 108, 110, 140–142
  cement, 1, 55, 59, 63
    replacement, 180
autoclave, 6, 140–41

bend over point, 71, 76, 98, 173
bending moment, 31
blastfurnace slag, 6, 7, 92, 102, 150–52
bloated clay, 107
bond
  fibre/cement, 49–54
  fibre/matrix, 35, 152
  frictional, 54
  glass/cement, 159
  interfacial, 38, 51, 49, 144
  strength, 157–61
bonding, 123
BOP *see* bend over point
bowing, 67
bundle solidification, 169

C–S–H *see* calcium silicate hydrate
$Ca(OH)_2$ *see* calcium hydroxide
calcium hydroxide, 6, 7, 92, 144, 147, 169
calcium silicate, autoclaved, 140
calcium silicate hydrate, 9, 20, 144, 152, 168
carbonation, 137, 139, 152, 164
catastrophic failure, 47
Cem-FIL, 4, 13
Cem-FIL 2, 13, 82
cement
  aqueous phase solution, 15
  blended, 7, 152, 168
  Chichibu, 9, 139
  extract, 3, 12, 17, 19, 166
  fineness, 6
  Fondu, 130
  Frodingham, 134, 138
  high-alkali, 7
  high-alumina, 5, 26, 129, 150–52
  hydraulic, 1, 5
  low-alkali, 2
  modified Portland, 92
  paste properties, 6
  Portland, 2, 5, 7
  Portland-blastfurnace, 5, 130, 139
  rapid hardening, 70
  regulated set, 5, 139
  Sealithor, 134
  sulphate resistant Portland, 70

Index

sulphoaluminate, 10, 139–40, 169, 182
    supersulphated, 5, 130, 134–8, 150–52
Cemsave, 102
cenospheres 6, 108, 109
Charpy impact strength, 127, 128
chemical
    extraction, 22
    inhibitors, 13
    resistance, 91
China clay waste, 92
chopped strand, 35, 56, 61
coating, 12–13, 170
compaction, 57
composites
    aligned, continuous, 28, 45, 172
    blended cement, 92–111
    high-alumina cement, 131–4
    lightweight, 107–11
    polymer impregnation, 129
    polymer modified, 112–29
    Portland cement, 70–91
    random 2-D, 35, 36, 41, 44, 54, 60, 143
    random 3-D, 34, 143
    sulphoaluminate cement, 139
    supersulphated cement, 134–8
compressive strength, 71, 76
concentric spray head, 60
Corning Glass Works, 3, 4
corrosion, 166, 174
crack
    path, 144, 161, 163
    size, 49
    spacing, 29, 30, 43, 45, 54, 160
    width, 37
cracking, 1, 102, 152, 161
    strain, 48–9
    stress, 48–9
crazing, 67
creep deformation, 90
critical
    energy, 28, 47
    fibre volume, 30, 31, 102, 173, 174, 176
    length, 35, 37, 38, 46, 54
    volume, 173–4
curing, 66, 124
    hot water, 82
    steam, 100

de-bonding, 49–53, 157, 172

de-watering, 58
density, 60, 68, 72, 88, 95, 113
diffusion, 20
dimensional changes, 88
drying shrinkage, 70
durability 5, 67, 76, 102, 109–10, 112, 133, 125, 164–76
dynamic frictional force, 41

efficiency factor
    length, 34–45
    orientation, 34–45
    strength, 38, 41
elastic modulus, 28, 34, 76
electron spectroscopy, (ESCA), 20
embrittlement, 147, 157, 165, 174
energy
    elastic, 46
    fracture, 45
    frictional, 46
    pull out, 46
    strain, 46
equivalent factor, 84, 86
    see also acceleration factor
ettringite, 137, 150, 152

failure strain, 30, 38, 71, 173
fatigue, 90
fibre
    alkali resistant glass, 1–25
    balling, 61
    bending failure, 161, 171
    bond strength, 160, 173
    bundle, 75, 147, 150, 157, 161, 167, 169, 174
    coating, 1, 11, 56
    composition, 11, 13
    content, 31, 63, 72, 82
    corrosion, 98
    diameter, 13, 40, 56
    dispersion, 143
    distribution, 68
    drawing temperature, 11
    E-glass, 1–5, 13, 26, 125–9, 133, 169
    embedment length, 51, 158, 160
    flaws, 2, 15, 22, 170
    G20 glass, 13, 17, 19, 20, 22, 166
        see also glass, G20
    glass, crystalline, 2
    glass ceramic, 2

glass rope, 1
inhibitive coating, 168
length, 13, 34, 72, 75, 82
manufacture, 10
mat, 41, 45, 63
orientation, 34, 68, 175
perimeter, 159–60
properties, 5
pull out, 37, 50, 132, 144, 147, 152, 154, 174
rovings, 1, 56, 63
slip, 28, 46
spacing, 49
strand, 11, 15, 63, 75, 84, 144, 158, 165, 168
strength, 5, 22–4, 167
viscosity, 5, 11
volume fraction, 26, 31, 49, 72, 76
Fibre Super, 13, 82, 168
filament(s), 11, 15, 17, 56, 75, 166, 174
filament winding, 62, 129, 143
fillers, 92, 93
fire
　protection, 60, 141
　resistance, 89
　tests, 89, 108, 109, 141
five/five pgrc mix, 124
Forton, 124, 125, 169
fracture surface, 98, 143–4
freeze–thaw, 91, 124, 164
Fuller's earth, 92

gamma irradiation, 129
gbfs *see* granulated blast furnace slag
glass
　composition, 3, 11
　G20, 3, 4
　　see also fibre(s), G20 glass
　interactions with cement, 17
　network breakdown, 12
　viscosity, 11
　zirconosilicate, 11, 12, 16, 166, 174
granulated blast furnace slag, 6, 7, 102, 154
GRCA, 5, 55, 69
grp, 1, 55
gypsum plaster, 134

Hatschek process, 55, 63–5
heat evolution, 94, 102
high-alumina cement
　composites, 130–34

conversion, 131, 133, 139

inhibitors, 13
injection moulding, 62
insulation, 60, 92, 108, 141, 182
interface
　fibre/cement, 133, 147, 154
　fibre/matrix, 49, 98, 144, 152
　glass/cement, 144
Italian pozzolana, 93
Izod impact strength, 71, 81, 105

Kieslguhr, 141

latently hydraulic materials, 168
lignosulphates, 56
limit of proportionality, 68, 71, 76, 93, 95, 108, 119, 132, 173
LOP *see* limit of proportionality

Magnani process, 55, 63
matrix
　cracking stress, 28
　cracks, 144, 161
　fibre stress, 28
　phase, 5–10
metakaolin, 107
methyl cellulose, 61
methyl methacrylate, 110
micro analysis, electron probe (EPMA), 21
microcracking, 163
microsilica, 6, 7, 105
microstructure, 143–63
microstructure of composites made from
　blended cements, 152, 154, 155
　high-alumina cement, 150, 153
　Portland cement, 144–50
　supersulphated cement, 152, 153
mixture rule, 28, 31, 34, 48
model composite, 28, 42
modulus of rupture, 30–31, 68
moisture movement, 67, 88
MOR/UTS ratio, 31, 42, 45
moulding, 62
multiple cracking, 28–31, 43, 45

National Research Development Corporation, 4
neutral axis, 31

Nippon Electric Glass, 177
nucleation, 174
   sites, 147

Owens Corning Fibreglass, 177
ozone, 124

perlite, 6, 107
permeability, 94, 102, 105, 168
permeance
   air, 89
   water vapour, 108
perpendicular strength, 108
pfa, 6–7, 55, 59, 92, 94–100, 113, 150–52
pigment, 56
Pilkington Brothers PLC, 4
pin gauges, 57, 68
plastic film, 67
plastic matrix theory, 54
Poisson's ratio, 71, 76
polyethylene oxide, 61
polymer(s), 2, 61
   dispersions, 6, 66, 112
   film, 113, 123, 169
   impregnation, 110–12, 129
   modification, 112–29, 169
polyvinyl acetate, 170
pore size distribution, 102
pore solution, 7, 130, 147, 168
porosity, 68, 75, 87, 94, 113, 129, 133, 147, 152
pozament, 95, 98, 155
pozzolana, 2, 7, 92, 93, 168
pre-mixed grc, 86, 87
processing methods, 61
production methods, 55–69
   asbestos cement, 63
   gravity moulding, 62
   injection moulding, 62
   lay-up process, 63
   manual spray, 57
   mechanical spray, 57
   miscellaneous spray, 59
   mix and place, 61
   spray de-watering, 58
   wet flat, 65
   winding process, 62, 63
pseudoductility, 75, 107, 163
pull-out
   grooves, 154
   load, 49, 51
   test, 52, 157
pumice, 107

quality control, 67
quarry fines, 92

rapid set, 92
refurbishment, 180
rendering, 59
respirability, 141–2
retarder, 56
rockwool, 140
roller compaction, 57

scanning electron microscope (SEM), 20, 143–63, 170
screw pulling, 108, 141
sea water resistance, 134
SEM *see* scanning electron microscope
shear mixing, 61
shear strength, 29, 172, 173
   in-plane, 72
   inter-laminar, 72, 76, 129
shipboard, 141
shrinkage, 66, 92
   drying, 2, 70, 88, 93, 124, 139
   ultimate, 88
silica fume, 7, 55, 105–107, 152
size, 12–13, 144, 170
sliding frictional force, 38, 41
slump, 68
sol-gel process, 11
spray-up process, 55–60
steam curing, 100
strain
   at first crack, 48
   energy, 30, 46
strand in cement test, 15, 16, 165
strength
   prediction, 84–6
   retention, 76, 123
stress
   at first crack, 48
   concentration, 170–72
   frictional, 50, 52
   reinforcement, 1
   rupture, 90
   shear, 49–52, 54

stress–strain
  curves, 27, 31, 38, 42, 45, 72, 102, 172
  model, 28
  response, 37, 38, 40, 41
styrene acrylonitrile, 129
styrene butadiene, 6, 123
sulphate resistance, 131, 134
superplasticiser, 105
surface finish, 67
swelling, 124

thermal
  conductivity, 88, 108
  expansion, 88
  insulation, 60, 89
tobermorite, 141
transverse sample, 100

ultimate tensile strength (UTS), 24
ultra-violet radiation, 124

vacuum forming, 59
vibration, 62, 87
vinyl propionate-vinyl chloride co-polymer, 123
volume fraction, 28

water absorption, 68, 124, 125, 129, 169
water:cement ratio, 6, 56, 58, 61, 87, 114, 125, 130
water:solids ratio, 68, 70, 98, 105
whiskers, 144, 147
work of fracture, 49
workability, 56, 61, 112
  aids, 56

X-ray fluorescence, 69

Young's modulus, 5, 23–4, 72, 81, 93, 107, 132